MOVING AMERICA TO METHANOL

# Moving America to METHANOL

## A Plan to Replace Oil Imports, Reduce Acid Rain, and Revitalize Our Domestic Economy

**Charles L. Gray, Jr.**
**Jeffrey A. Alson**

**THE UNIVERSITY OF MICHIGAN PRESS**
Ann Arbor

Published in the United States of America by
The University of Michigan Press and simultaneously
in Rexdale, Canada, by John Wiley & Sons Canada, Limited
Manufactured in the United States of America

1988 1987 1986 1985 4 3 2 1

*The views expressed herein are those of the authors and do not necessarily represent the views of the United States Environmental Protection Agency.*

**Library of Congress Cataloging in Publication Data**
Gray, Charles L., 1946–
Moving America to methanol.

Bibliography: p.
1. Methanol fuel industry—Government policy—
United States. I. Alson, Jeffrey A., 1956–
II. Title.
HD9502.5.M473U63 1985 338.4'7665772 85-16469
ISBN 0-472-10071-8
ISBN 0-472-08063-6 (pbk.)

# Acknowledgment

The authors would like to express their appreciation to Vivian Wallace for her assistance in preparing the manuscript for this book. Her dedication and enthusiasm survived numerous revisions.

# Contents

# Introduction

Support continues to grow for the suggestion that acid rain is causing significant environmental problems in several areas of North America. Acid rain consists of sulfuric and nitric acids. The sulfuric acid is formed primarily from the combustion of high-sulfur coal in utility power plants. Nitric acid also is largely due to stationary fuel combustion sources, but motor vehicles contribute to nitric acid levels as well. Most of the concern over the effects of acid rain is focused on the northeastern United States and eastern Canada. Most scientists believe that the impact of the sulfuric acid component of acid rain in these areas is more important than the impact of nitric acid. Further, most scientists agree that the principal sources of sulfuric acid in these areas are utility power plants in the Midwest and East burning high-sulfur coal.

Many specific proposals to reduce sulfur dioxide ($SO_2$) emissions (which result in the formation of sulfuric acid in the atmosphere) from utility power plants have been suggested. Proposed solutions differ in detail, but generally involve one of two basic approaches. One strategy is the establishment of a performance standard, an emission standard that limits the amount of $SO_2$ that can be emitted from a given size power plant. This approach would rely on the marketplace to decide the most economical means of achieving the standard.

The second strategy is to require the installation of a specific technology (called flue gas desulfurization or "scrubbers") on utility power plants to remove $SO_2$ from the exhaust gases. This approach has developed as an alternative because utilities in large part would use western low-sulfur coal, which produces less $SO_2$, in place of currently used eastern and midwestern high-sulfur coal as the lowest cost means of complying with a performance standard. The consequence of using low-sulfur coal is that for large $SO_2$ reductions (on the order of 8 to 10 million tons per year) over 100 million tons per year of eastern and midwestern coal would likely be displaced. This would be a severe economic blow to these coal-producing areas.

Those who argue for the required installation of scrubbers are doing so to protect the existing market for high-sulfur coal, since it is generally accepted that mandating scrubbers for a large $SO_2$ control program would cost an additional $10 billion to $30 billion over and above the cost of a performance standard. We suggest that the appropriate question is not "Should the current market for high-sulfur coal be maintained?" but rather "What is the most appropriate use of this country's vast reserves of high-sulfur coal?"

The U.S. energy problem is one of petroleum supply. The primary sources of energy in the United States are petroleum, natural gas, coal, hydropower, and nuclear power. Except for petroleum, the United States has a very good supply of each of these energy sources. Petroleum, however, is our most important fuel and currently accounts for approximately 40 percent of our total energy consumption. The United States must, because of limited reserves, import a large proportion of its petroleum needs. Figure 1 shows a projection of U.S. oil supply to the year 2000. Although the economic recession has contributed to a reduction in oil consumed in the United States since 1980, the total net import bill for oil is still quite substantial (approximately $50 billion in 1984). While future cost projections are very uncertain, the net oil import bill could easily exceed $100 billion per year before the end of this decade even assuming no world oil supply disruptions. Supply restrictions could significantly increase world oil prices and oil import payments.

Over one-half of all U.S. petroleum consumption today is in the transportation sector. In the future, since coal, natural gas, and electricity can be more easily substituted for oil in other sectors of the economy, transportation can be expected to account for an even larger share of U.S. petroleum con-

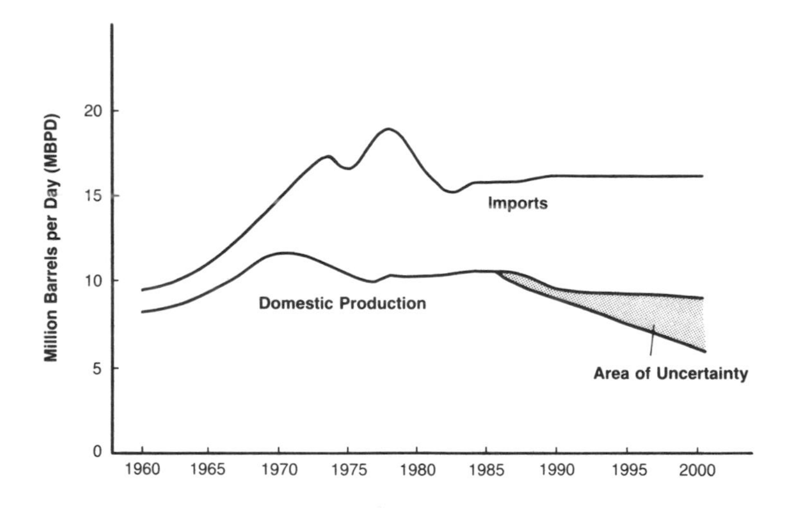


**Fig. 1.** Past and projected U.S. oil supplies. High and low projections are shown for domestic oil production after 1985, which result in high and low values for levels of future oil imports as well.

sumption. Therefore, the key to effectively dealing with our "energy problem" is the identification of and transition to an alternative transportation fuel which can be produced from domestic resources.

## Solving the Problems

An optimum solution exists to both the problem of acid rain and the problem of U.S. dependence on imported petroleum. Burning low-sulfur coal in utility power plants and converting the displaced high-sulfur coal to methanol for transportation would address the two problems simultaneously, providing some beneficial synergistic effects. For example, the excess capital ($10 billion to $30 billion) that would have been invested in scrubbers to reduce $SO_2$ emissions is no longer needed, or could be viewed as being available to offset a significant portion of the capital required to construct coal-to-methanol plants and thereby reduce the cost of methanol to the consumer. While the idea is rather simple in concept, the government action that would be required to ensure that the objective is achieved is not so simple.

There is a consensus among transportation energy experts that methanol is the motor vehicle fuel of the future. Methanol has long been recognized as a highly efficient motor vehicle fuel. Methanol vehicles have very low emissions and are therefore environmentally attractive. Further, methanol can be produced from our largest domestic energy resource, coal, utilizing production technologies that are available today. The transportation fuel market can easily accommodate large volumes of methanol fuel. Even if 130 million tons of high-sulfur coal (the amount of coal that could be displaced by a major acid rain control program that permitted switching to low-sulfur coal) were converted to methanol annually, this methanol would supply less than one-seventh of U.S. highway fuel demand. Once the methanol transition gathered momentum, a clear potential exists for growth in high-sulfur coal demand.

Of course, critical to the marketability of methanol fuel produced from coal is its relative economics as compared to gasoline from petroleum. The retail price of methanol from coal is projected to be equivalent on an energy basis to petroleum at approximately $43 per barrel (all monetary values are in 1982 dollars except for historical values that are given in current dollars) and could easily be competitive with gasoline from imported petroleum by the 1990s. However, methanol-fueled engines are expected to be 20 to 30 percent more efficient than competing gasoline engines. Therefore, from a consumer standpoint, methanol is likely to be economically attractive when compared to gasoline from oil costing more than about $35 per barrel. Considering the long lead times associated with a transition to an alternative transportation fuel, major efforts should be under way now if a significant supply of methanol for transportation is to be available by the mid or even late 1990s.

If methanol for transportation is such an excellent idea, why isn't the transition occurring? The basic problem is simply the inertia that propels the existing petroleum-based transportation system. It would be exceedingly difficult for any alternative fuel not already conveniently available to the motorist to break into the transportation market. The problem has often been referred to as a classic "chicken-and-egg" dilemma. Fuel suppliers cannot afford to invest the extremely large capital outlays required for coal-to-methanol plants without certainty that a market for the methanol will exist, and the producers of motor vehicles cannot commit to market methanol-fueled vehicles without assurance that the fuel will be conveniently available to motorists during the model year of introduction. Most who have studied this issue have concluded that some form of government coordination will be necessary for an efficient and nondisruptive transition. The question that remains is, "What specific government action is needed?"

### The Implementation Plan

Clearly the economic, energy, and environmental benefits of methanol's use as an automotive fuel are so strong that ultimately the marketplace will dictate its use. Some suggest that all government need do is remove any regulatory or institutional barriers that might inhibit the transition to methanol. But even if all regulatory uncertainties or barriers to methanol were removed, there would still be the critical implementation questions concerning the timing and magnitude of private sector investments that will unnecessarily delay the use of methanol. These are the true implementation barriers to the methanol transition.

We believe that now is the time to begin the methanol transition and that incentives to produce and supply methanol from high-sulfur coal are necessary, as well as incentives to ensure methanol fuel demand. The incentives we recommend to stimulate production of methanol are: (1) a highway fuel tax, (2) a price guarantee for methanol from high-sulfur coal, and (3) a requirement for methanol availability at certain service stations. The incentives to ensure methanol fuel demand are: (1) a tax credit for methanol vehicle purchases, (2) a requirement that new urban buses use methanol, and (3) a requirement that new federal vehicles be methanol-fueled. The details of these initiatives were carefully designed to minimize adverse impacts and to provide for a minimal level of governmental involvement.

The two primary effects of this program will be to reduce acid rain and lessen U.S. dependence on imported petroleum. Sulfur dioxide emissions, and thus acid rain levels, will be significantly reduced as a direct consequence of using low-sulfur coal in utility power plants. Less petroleum will need to be imported as a direct result of substituting methanol for petroleum in the transportation sector. In addition, implementation of this program will also:

1. lower the overall cost of electricity
2. provide a major new market for low-sulfur coal
3. maintain and ultimately expand current demand for high-sulfur coal
4. improve the balance of trade
5. increase economic growth and employment
6. provide insurance against inflation
7. improve national security
8. reduce the federal deficit
9. result in a U.S. lead in methanol technologies
10. improve atmospheric visibility
11. encourage development of vehicles with low emissions and high efficiencies
12. help stabilize world oil prices.

Consumers will experience the lowest overall electricity cost possible under an acid rain control program since the marketplace will be allowed to achieve the desired $SO_2$ reductions in the most cost-effective manner possible. With utility power plants switching from high-sulfur coal to low-sulfur coal a major new market for low-sulfur coal will rapidly develop. Since this program is designed to provide methanol from high-sulfur coal, the existing demand for high-sulfur coal will be assured, with expansion of demand for high-sulfur coal expected when petroleum prices rise in the future.

The U.S. balance of trade would significantly improve as a result of several factors. The most direct effect will be reduced payments for imported oil, on the order of $20 billion to $30 billion per year by the mid-1990s, and possibly higher thereafter. Secondary effects would appear in the following areas: improved competitiveness of the U.S. petrochemical industry due to lower petroleum prices, demand for U.S. methanol production technologies, and reduced levels of imported automobiles as the U.S. automobile industry takes the lead in methanol-fueled vehicle production technology.

As a direct result of producing even the minimal methanol fuel stimulated by this program, over a quarter million new U.S. jobs would be created in the late 1980s and early 1990s, with the peak level of new jobs reaching 400,000. A methanol transportation fuel program would also have a constraining effect on inflation since, once the investment is made in a coal-to-methanol plant, the price of methanol will be relatively insensitive to inflationary pressures. Forty to 60 percent of the cost of methanol from a coal-to-methanol plant will be associated with the cost of the capital to construct the plant, depending upon financing methods. Once the capital investment is made, labor and coal costs become the two significant factors in the price of methanol. Since coal is so readily available in the United States and the cost of labor should only go up in proportion to overall inflation, future increases in methanol fuel price would be modest and, in constant dollars, could actually go down. In addition, reducing

U.S. dependence on imported oil would lessen the impact of future world oil price rises on the U.S. economy.

The United States and most of the Western world are critically dependent upon petroleum from some politically unstable parts of the world. By reducing dependence on imported petroleum and providing an alternative for other countries as well, national security would be directly improved.

This program would reduce the federal deficit both directly through the highway fuel tax and indirectly through economic growth. Increased federal revenues from the highway fuel tax should be between $20 billion and $30 billion per year in the 1990s. Additional revenues would be available from taxes on the wages from the hundreds of thousands of new jobs created under this program, and the demand for social welfare spending should be lessened. This program would be a significant step toward reducing the federal budget deficit.

Feedstocks for methanol production are available in large quantities throughout the world. Many countries with little or no petroleum reserves have resources that could be used to produce methanol. The United States, in developing and refining the technologies of both methanol production and methanol-fueled vehicles, would be in a unique position to provide technological assistance to other countries.

Reduction of $SO_2$ emissions would not only reduce acid rain but would reduce the formation of sulfates that impair atmospheric visibility and have been associated with various health effects. In addition, to the extent that methanol fuel displaces diesel fuel, urban visibility would improve as diesel particulate levels are reduced. Methanol-fueled vehicles will have lower overall emissions as well as higher engine efficiencies.

The move to a methanol transportation fuel system in the United States would have a stabilizing influence on world oil prices. In choosing a highway fuel tax, the United States will be setting a cap on world oil prices. Barring artificial constraints in the marketplace and considering the large share of world oil consumed by this country, the United States is in a unique position to influence world oil prices for at least the next several years when petroleum supplies will theoretically still be sufficient to meet demand. If methanol provides a real alternative to petroleum as a transportation fuel, petroleum-producing countries must realize increasing oil prices will stimulate additional investment in coal-to-methanol plants in the United States. This investment would encourage an irreversible commitment to a replacement for their product, oil.

This book addresses many of the most important issues facing the United States today, several of which are expected to intensify in the future. We believe one of the strongest attributes of this book is that it recognizes the direct and indirect relationships between the environmental problem of acid rain, the energy problem of dependence on imported petroleum, and economic problems such as growth, trade imbalances, and budget deficits. While each of these issues has been the subject of numerous studies, this book is a unique attempt to consider them within one analytical framework and to propose an integrated and coordinated strategy to solve or begin to solve all of these problems.

From a purely political perspective, it might at first appear that this is a very inappropriate time to advocate such a comprehensive methanol implementation program. The political furor that had characterized the acid rain debate has subsided and world oil prices continue to fall slowly but steadily. But beneath the calm lie several unsettling realities: scientific concern over the impacts of acid rain continue to deepen, both domestically and internationally; oil import levels are rising, and the increases will undoubtedly accelerate given reduced conservation and domestic oil exploration; our monthly trade deficits continue to set records; progress in reducing the federal budget deficit has stalled; and economic growth has slowed much sooner than expected.

One primary reason why these issues have not yet been addressed is the political perception that they have not become crises. In fact, the noncrisis atmosphere offers an excellent opportunity for sound and creative public policies. Waiting to resolve a problem until it has reached crisis proportions (such as a third oil price shock, for example) increases the likelihood that political, and not technical, considerations will dictate the policy debate. Recent events suggest that the time is right for many of the strategies proposed in this book. For example, the optimum time to enact a petroleum tax is during a period of falling world oil prices; the addition of a $0.40 per gallon highway fuel tax would yield overall gasoline prices only slightly higher than those that occurred in 1981. Congressional scrutiny over the fate of the Synthetic Fuels Corporation provides an excellent opportunity to consider a targeted coal-to-methanol price guarantee program. With lead being rapidly phased out of gasoline, and with diesel car sales plummeting, there is considerable flexibility for service stations to utilize existing storage facilities for a new vehicle fuel like methanol.

In summary, *now* is the time for implementing a program that will coordinate the transition to methanol as a national transportation fuel and that will contribute to the resolution of many critical problems facing our country. We hope this book will advance this important process.

# Part 1 The Acid Rain and Energy Problems—The Optimum Solution

# CHAPTER 1 The Problems of Acid Rain and Imported Petroleum

Acid rain is one of the most significant environmental issues of the 1980s. The debate over the appropriate public policy response to acid rain has reached a status normally achieved only by national security and macroeconomic policy issues. The national consensus that developed in the latter part of the 1970s on environmental protection policy has fractured as various powerful interest groups have spurned compromise and drawn the battle lines. Environmental advocates argue that acid rain is leading to irreversible ecological damage and must be halted regardless of the economic cost. Industry representatives argue that the ecological fears are overstated and/or unproven and that the economic burden of control is far too great. The acid rain debate is further complicated by strong regional interests. Residents and political representatives of the Northeast believe that their lakes, streams, and forests, so critical to their economic well-being and widely regarded as national treasures, may be irreversibly damaged or destroyed by pollution for which they are not responsible and over which they have no control. The Midwest, home of much of the U.S. industrial base and only recently recovering from several years of economic depression, is faced with the frightening prospect of control programs that could increase electrical rates or cause massive reductions in coal mining and related employment. Acid rain also has international implications, as Canada is a major recipient of U.S. emissions that lead to acid rain and has expressed considerable frustration with the lack of U.S. action to reduce these emissions.

In view of these conflicts, it is not surprising that acid rain has become a major political issue. Environmental groups and northeastern leaders consider acid rain controls to be a top legislative priority, while utilities, high-sulfur coal owners and miners, and midwestern representatives focus their efforts on defeating or delaying controls. Public opinion polls have also indicated that acid rain is an important issue to a significant portion of the general populace.

The science of acid rain can be divided into two broad areas: causes and effects. That precipitation is in fact acidic over large areas of the United States (and the North American continent) is certain. The acidity or alkalinity of a solution is measured on the pH scale. A solution that is neutral (neither acidic nor alkaline) has a pH of 7.0. A pH below 7.0 indicates an acidic solution; a pH above 7.0 indicates an alkaline solution. The natural acidity of rainwater is generally assumed to be between pH 5.6 and pH 5.7, the value for pure water in equilibrium with atmospheric concentrations of carbon dioxide ($CO_2$). Naturally occurring chemical species other than $CO_2$ can also affect rainwater pH, however, and it is probably more appropriate to consider historical pH values for rain to be between 4.9 and 6.5.[1] Monitoring data indicate that nearly the entire United States east of the Mississippi River (excluding the deep South) receives precipitation with an annual mean pH of 4.4 or less.[2] Since the pH scale is logarithmic, such precipitation is three to one hundred times more acidic than historical values. Moreover, these regions also receive sulfates and nitrates deposited in dry form. Though dry deposition has been poorly monitored, it is believed that dry deposition is nearly as important as wet deposition in the overall acidification phenomenon.[3]

An ecosystem may or may not be damaged by acidic deposition depending upon its alkalinity, or acid neutralizing capability. Studies have shown that the following regions contain aquatic systems sensitive to acidification: nearly all of New England and eastern Canada, much of the Allegheny, Smoky, and Rocky mountain areas, and parts of the north central and northwestern United States.[4] Studies of individual lakes in New England and eastern Canada show increasing acidification over the last few decades, and there is considerable evidence of damage to aquatic life. That acidification of surface waters can lead to lower fish populations has been documented in the Adirondacks, Ontario, and in Scandinavia.[5] Aquatic life can be disrupted by a number of acid-related processes, but direct acid toxicity and toxic metal poisoning are probably the most common mechanisms. The outlook for the future is particularly bleak. Under current rates of acidic deposition, a long-term pH of 4.9 or less can be expected for low-alkalinity lakes and streams in New England and southeastern Canada. At this pH, most fish species, virtually all mollusks, and many forms of lower aquatic life will be eliminated. Those fish species that can tolerate the low pH may be eliminated by high levels of aluminum leached from nearby soils.[6] Thus, unless acid deposition rates are lowered, it is possible that many lakes and streams in the northeastern United States and southeastern Canada will lose most aquatic life within a few years to a few decades.

While the most documented impacts of acidification have been on aquatic life, there are several other areas of concern as well. Although there is as yet no definitive evidence that acidic deposition impacts forest growth, there is strong circumstantial evidence involving widespread reports of decreased tree growth and increased tree mortality in several regions with high acid deposition.[7] Some agricultural crops have been shown to be negatively affected by acidic deposition, but other crops seem to be positively affected while a majority of crops are not affected at all.[8] Continued research should lead to a resolution of these issues. If a link between acidic deposition and reduced tree or crop growth is found, the economic consequences of acid rain would be greatly increased.

Finally, there is the question of indirect human health effects of acidic deposition due to increased metal mobilization. The two pathways of concern are bioaccumulation of toxic chemicals in the human food chain and the toxic contamination of drinking water. Methyl mercury has a low toxicity to fish but is highly toxic to the human central nervous system. It is produced from inorganic mercury by bacteria present in water sediments and is highly concentrated in certain fish, especially predatory freshwater fish.[9] Fish consumption is nearly the only exposure pathway for methyl mercury to man.[10] Many studies have found elevated mercury levels in fish to be associated with high levels of acidity, though a definite cause-effect relationship has not been demonstrated.[11] Acid mobilization of toxic metals is a concern for all drinking water supplies, as it can liberate metals such as mercury, cadmium, and aluminum in a watershed and can corrode metals such as lead and copper out of plumbing systems. High levels of lead and copper have been found in the water of homes utilizing rainwater cisterns as well as in cities with lead water distribution pipes.[12] Lead, of course, is an important human neurotoxin that has particularly serious effects on children. The reduction of lead exposure is a national goal and any increased exposure due to water acidity must be viewed as an incremental burden. While it is not possible to quantify the impacts of toxic metal mobilization on human health at this time, it must be emphasized that there has been little research directed at these issues to date. Both fish bioaccumulation of toxics and contamination of drinking water supplies may ultimately prove to be important impacts of acid deposition.

While there has been general agreement as to the occurrence and effects (at least with respect to aquatic damage) of acid rain, there has been considerable controversy over the causes of acidic deposition in eastern North America. Certain facts cannot be contested, however. Acidity is caused by increased concentrations of the hydrogen ion, which in turn are increased by higher concentrations of the sulfate ion ($SO_4$) and the nitrate ion ($NO_3$). Acid rain refers to precipitation containing high levels of sulfuric ($H_2SO_4$) and nitric ($HNO_3$) acids, which are the forms that sulfate and nitrate ions take in solution. The sulfur and nitrogen compounds that are ultimately transformed to sulfate and nitrate result from both natural and man-made sources, but the latter are predominant in the United States. Approximately 99 percent of all sulfur and nitrogen emissions east of the Mississippi River are from man-made activities.[13] Sulfates are considered to be responsible for a majority of the acidity in the eastern United States. Approximately 70 percent of all sulfur oxide emissions east of the Mississippi River are from electric utilities, with nearly all of these emissions from coal-fired plants.[14] Furthermore, the largest sources of sulfur oxide emissions tend to be concentrated in certain parts of the Midwest. Specifically, seven states in the Ohio River valley (Ohio, Pennsylvania, West Virginia, Indiana, Kentucky, Illinois, and Missouri) accounted for over 50 percent of all sulfur oxide emissions in 1980 in the thirty-one-state region to the east of and bordering the Mississippi River.[15] Nitrogen oxide emissions in the eastern United States are nearly evenly divided between electric utilities, motor vehicles, and other miscellaneous sources.[16]

The popular explanation for the cause of acid rain in the northeastern United States is as follows: high levels of sulfur dioxide emissions from midwestern power plants (primarily in the Ohio River valley), with associated nitrogen oxide emissions, are released through very tall smokestacks; these gases are transported eastward and northward hundreds of miles by prevailing winds and in the process are converted to sulfate and nitrate ions; ultimately these sulfate and nitrate ions are deposited in the northeastern United States or southeastern Canada as acid rain or as dry acid particles. It is this explanation which has generated considerable controversy, as it includes several assumptions that have not been empirically verified. One notable debate has concerned the question of linearity, i.e., whether there is a direct linear relationship between precursor emissions and acidic deposition or whether there is a limiting factor that causes a nonlinear relationship. A second important issue is the relative influence of local versus distant sources in causing the acid deposition problems in the northeastern United States and southeastern Canada. These issues have been central to the controversy over acid rain because the concepts of linearity and long-distance transport form the foundation of virtually all of the major proposed acid rain control programs. Thus, much of the recent scientific discussion has addressed these issues.

A scientific consensus has begun to emerge on the issue of acid rain. Evidence can be found in the reports of two influential acid rain research groups. In the fall of 1983 the National Academy of Sciences released its long awaited report on atmospheric transport. After an exhaustive study of the pertinent data, the academy's Committee on Atmospheric Transport and Chemical Transformation in Acid Precipitation concluded that reductions in sulfur oxide emissions in eastern North America would result in lower acid deposition rates in eastern North America, though the committee stated that it would not be possible to predict the impacts on specific receptor regions with a high degree of certainty. The report found that "there is no evidence for a strong nonlinearity in the relationships between long-term average emissions and deposition."[17] Thus, a percentage reduction in sulfur oxide emissions would result in a similar percentage reduction in acid deposition. The committee also found no evidence that either local sources or distant sources were dominant in terms of acidic deposition.[18] In July 1984, the White House Office of Science and Technology Policy acid rain review panel issued its final report. It stated that reducing sulfur dioxide emission levels would reduce total deposition levels and, accordingly, lower the probability that major changes would occur in sensitive receptor regions. With respect to policy decisions in the midst of scientific uncertainty, the panel stated that "it is in the nature of the acid deposition problem that actions have to be taken despite incomplete knowledge. . . . Recommendations based on imperfect data run the risk of being in error; recommendations for inaction pending collection of all the desirable data entail the even greater risk of ecological damage."[19] The panel was particularly concerned about the possibility of irreversible environmental effects which could transpire in the ten to twenty years it would take to complete acid rain research and advocated "meaningful reductions in the emissions of sulfur and nitrogen compounds."

The importance of the National Academy of Sciences and Office of Science and Technology Policy reports, one issued by a committee from the most prestigious scientific organization in the country and the second by a panel appointed by the White House, cannot be underestimated. The conclusions of the two reports have altered the public and political views of the science of acid rain. Although both reports emphasize the uncertainties involved in their analyses, the inescapable scientific consensus represented by these reports is that an acid rain control program must be undertaken in the near future if we are to avoid the risk of major irreversible environmental impacts and that the most appropriate control program would involve sulfur oxide emission reductions in the region east of the Mississippi River.

Translating this emerging scientific consensus into a political solution has proven to be very difficult. A plethora of acid rain control proposals have been introduced in Congress in recent years. Acid rain legislation must address several issues: the magnitude and timing of the sulfur dioxide reductions; identification of the area targeted for reductions and a mechanism for apportioning those reductions within the area; whether or not to include an emission cap or emission offsets for future growth; whether to allow intrastate, interstate, or nitrogen oxide/sulfur oxide trading; whether to allow sources complete flexibility in choosing how to achieve emission reductions or whether to mandate control technology; and how to finance the control program. All of the major acid rain control proposals call for reductions of between 8 and 12 million tons of sulfur dioxide by sometime in the 1990s. Some pieces of legislation call for caps, offsets, and trading, while others do not, but these concepts can be added to or deleted from any proposal without altering its fundamental design. The basic issues that distinguish the most important acid rain control program proposals are the questions of whether or not control technology should be mandated, and financing. These differences are exemplified in two recent congressional proposals. One Senate bill, sponsored by Senators Stafford, Chafee, and Mitchell, would require a 10-million-ton sulfur oxides reduction in the thirty-one-state area east of and bordering the Mississippi River. It would allow local sources to determine how best to meet the mandated reductions, thus allowing, in effect, a "least-cost" solution. Local sources would also have to pay for the reductions, resulting in higher electricity rates for utilities reducing emissions. A popular House resolution, introduced by Representatives Waxman, Sikorski, and Gregg, would require a 10-million-ton sulfur oxides reduction (and nitrogen oxide reductions as well) over the continental forty-eight states. A reduction of 7 million tons would be achieved by mandating the installation of flue gas desulfurization devices (scrubbers) on fifty of the largest utility power plants, with the remaining 3 million tons of reductions delegated to the states. Ninety percent of the cost of installing scrubbers would be reimbursed by the government from a fund generated by a 1 mill per kilowatt-hour tax on all nonnuclear electricity.

The question of whether to allow least-cost controls or mandate scrubbers is a particularly volatile debate. The concept of least cost recognizes that generally regulations should permit compliance in as economically efficient a manner as possible. It is widely recognized that in most cases it would be cheaper for midwestern utilities to switch from high-sulfur coal to low-sulfur coal than to retrofit scrubbers. The problem is that if a large number of utilities were to switch to low-sulfur coal, the demand for high-sulfur coal would drop significantly and a large number of coal miners would be put out of work. It is this reality that has driven support for the scrubber mandate which would allow high-sulfur coal to continue to be used. However, analyses of such proposals have indicated that a scrubber mandate would raise total capital costs of an acid rain control program by approximately $20 billion and would increase the total cost of control by approximately $1 billion per year.[20]

The battle lines in Congress have been drawn along regional lines, as expected, but the problem has been exacerbated by the compositions of the House and Senate committees which have responsibility for the Clean Air Act. The Senate Environment and Public Works Committee has several members from the Northeast and none from the Ohio River valley, so it would be expected to favor a least-cost, thirty-one-state, polluter-pays approach. Such an approach would also be expected to be popular on the Senate floor, as it would allow western senators to support a control program without impacting utility rates in their states. The House Energy and Commerce Committee, on the other hand, has heavy representation from the industrial Midwest. To date it has been deadlocked over acid rain legislation. But if the House Energy Committee approves any program, it is likely to be a scrubber-mandated, forty-eight-state, cost-sharing approach. Thus, politics dictate that the House and Senate are likely to support widely differing bills. This fact has resulted in a policy paralysis on acid rain, despite the growing consensus in Congress that a control program needs to be adopted. The search for a politically viable program, which significantly reduces sulfur dioxide emissions in a cost-effective and nondisruptive way, continues.

## Imported Petroleum

It was just a little over a decade ago that energy security became a national priority. Until then American society had prospered on cheap and available energy supplies, and it was only after the oil embargo by the Organization of Petroleum Exporting Countries (OPEC) in 1973–74 that we recognized our vulnerability with respect to energy security. Concerns over energy policy were reinforced in 1979–80 when the relatively minor cutoff of crude oil from Iran caused major havoc with world oil prices. Political interest in energy issues tends to be rather cyclic. When oil prices are on an upward swing or when market supplies begin to tighten, political leaders take note. Alternately, when oil prices are stable or falling and supplies are abundant, then little attention is paid to energy. There is increasing recognition within the technical community, however, that long-term energy policy is a serious issue that must not be addressed only in short-term crisis situations.

Central to an understanding of our energy situation is the realization that we do not have an energy problem per se, rather, we have a petroleum supply problem. Our present sources of energy, in order of importance, are petroleum, natural gas, coal, hydropower, and nuclear power. Except for petroleum, we have adequate domestic supplies of each of these fuels. Petroleum, however, is our most valuable fuel and has been a critical factor in post–World War II industrial expansion. Petroleum currently accounts for 43 percent of our total energy consumption.[21] Because it is a liquid and can be refined to several useful products, petroleum is a premium energy source. In the form of gasoline, diesel fuel, and jet fuel, petroleum is the foundation for our entire transportation system. As distillate fuel oil, residual fuel oil, and assorted specialized fuels, petroleum is also used for residential and commercial space heating, as an industrial and utility boiler fuel, and is an important petrochemical feedstock.

Unfortunately, U.S. demand for petroleum far outstrips domestic production. Table 1 shows overall U.S. petroleum disposition since 1972, broken down into domestic production, imports, exports, and total petroleum consumption.[22] It can be seen that total petroleum consumption has averaged around 17 million barrels per day (MBPD), or 6.2 billion barrels per year, over the last decade. With total proven reserves of crude oil and natural gas liquids in the United States of 34.6 billion barrels, if we relied solely on

TABLE 1. U.S. Petroleum Supplies from 1972 to 1984 (MBPD)

| Year | Domestic Production | Imports | Exports | Total Petroleum Consumption |
|---|---|---|---|---|
| 1972 | 11.18 | 4.74 | .22 | 16.37 |
| 1973 | 10.98 | 6.26 | .23 | 17.31 |
| 1974 | 10.50 | 6.11 | .22 | 16.65 |
| 1975 | 10.05 | 6.06 | .21 | 16.32 |
| 1976 | 9.77 | 7.31 | .22 | 17.46 |
| 1977 | 9.91 | 8.81 | .24 | 18.43 |
| 1978 | 10.33 | 8.36 | .36 | 18.85 |
| 1979 | 10.18 | 8.46 | .47 | 18.51 |
| 1980 | 10.21 | 6.91 | .54 | 17.06 |
| 1981 | 10.23 | 6.00 | .60 | 16.06 |
| 1982 | 10.25 | 5.11 | .82 | 15.30 |
| 1983 | 10.30 | 5.05 | .74 | 15.23 |
| 1984 | 10.44 | 5.38 | .72 | 15.71 |

*Note:* Generally petroleum consumption equals domestic production plus imports minus exports, but values will not be exact due to processing gains and losses, change in stocks, round off, etc.

domestic reserves we would run the risk of depleting those reserves within a decade.[23] Estimates of undiscovered recoverable resources of petroleum in the United States vary. In 1981 the U.S. Geological Survey (USGS) estimated that there were 82.6 billion barrels of undiscovered recoverable petroleum in the United States, though a more recent USGS report has stated that that estimate may be too high based on recent drilling experience.[24] Even assuming the 1981 USGS estimate is correct, the sum of the proven reserves and undiscovered recoverable resources would be just 117.2 billion barrels, which would last only nineteen years at an average consumption of 6.2 billion barrels per year.

Accordingly, as table 1 shows, the United States has been importing significant volumes of petroleum for many years. Table 2 shows the levels of net oil imports (imports minus exports) since 1972, along with the cost of imported oil and the total U.S. net oil import bills.[25] Net oil imports were 4.5 MBPD in 1972, rose to a peak of 8.6 MBPD in 1977, and dropped to 4.3 MBPD in 1982 and 1983. The net cost of imported oil was just $5 billion in 1972, rose to $79 billion in 1980, and was $46 billion in 1983. Both of these recent trends reversed in 1984. Net oil imports rose to 4.7 MBPD and our net import bill rose to just under $50 billion. Table 2 shows that the U.S. oil import bill has fallen since 1980 primarily because our net oil import levels are approximately half of what they were in 1977. As can be seen in table 1, however, domestic production has increased only slightly since 1979. Thus, our large import reductions are a direct reflection of our lower overall petroleum consumption since 1978, and are not a result of increased domestic production.

TABLE 2. Cost of Net Crude Oil and Petroleum Products Imports to the United States from 1972 to 1984

| Year | Net Imports (MBPD) | Refiner Acquisition Cost of Imported Crude Oil ($ per barrel) | Total U.S. Net Import Bill ($ billion) |
|---|---|---|---|
| 1972 | 4.52 | 2.90 | 4.8 |
| 1973 | 6.03 | 3.70 | 8.1 |
| 1974 | 5.89 | 12.55 | 27.0 |
| 1975 | 5.85 | 12.30 | 26.3 |
| 1976 | 7.09 | 13.48 | 34.9 |
| 1977 | 8.57 | 14.53 | 45.5 |
| 1978 | 8.00 | 14.57 | 42.5 |
| 1979 | 7.99 | 21.67 | 63.2 |
| 1980 | 6.37 | 33.89 | 78.8 |
| 1981 | 5.40 | 37.05 | 73.3 |
| 1982 | 4.30 | 33.55 | 52.7 |
| 1983 | 4.31 | 29.30 | 46.1 |
| 1984 | 4.66 | 28.88 | 49.1 |

*Note:* Net imports equals total imports minus total exports.

There have been two underlying causes of the lower U.S. petroleum consumption since 1978: low domestic economic growth and price-induced conservation. With world oil prices falling since 1981, conservation efforts have recently slowed. Expectations are that as the domestic economy continues to recover, total petroleum consumption will begin to rise. In addition, most experts project that domestic production, which peaked in 1970, will gradually begin to fall again in the late 1980s. Based on estimates from the Department of Energy (DOE), the Office of Technology Assessment, Exxon, and Conoco, it is reasonable to assume that domestic petroleum production will be in the range of 7 to 9 MBPD during the early 1990s.[26] Total petroleum demand projections by these same authorities for this time frame range from 14 to 16 MBPD. Thus, it is clear that barring the rapid introduction of alternative fuels and/or economic conditions so severe as to drastically curtail demand, oil imports will increase well above current levels. Projections are that imports will range between 5 and 9 MBPD during the early 1990s.

What will these future oil imports cost? Table 2 shows that imported crude oil cost peaked in 1981 at an annual average of $37 per barrel. Since early 1981, the world price of oil has been slowly but steadily dropping and averaged $29 per barrel for U.S. imports in 1983 and 1984. Predicting future world oil prices is very difficult, of course, due both to the possibility of a major conflict in the Middle East that could disrupt supplies and to uncertainties surrounding OPEC cohesion, non-OPEC oil discoveries, economic growth and conservation in the oil-consuming countries, etc. Most analysts believe that oil prices will remain stable or possibly fall slightly in real dollars in the near term but will likely rise in real terms in the late 1980s and 1990s.

Assuming no major oil supply disruptions, DOE has estimated that world oil prices will range from $26 to $40 per barrel in 1990 and $31 to $60 per barrel in 1995.[27] These values are all expressed in constant 1982 dollars. Based on these projected prices and a range for net imports of 5 to 9 MBPD, table 3 yields some broad ranges for future costs to the United States for imported oil. It can be seen that we could be paying over $100 billion annually for imported oil by 1990 and nearly $200 billion annually by 1995. Because of the interdependence among world oil price, oil demand, and domestic production, it would be most reasonable to assume low levels of imports at the highest oil prices and high levels of imports at the lowest prices. Doing so would indicate that the U.S. bill for imported crude would likely be between $73 billion and $85 billion in 1990 and between $102 billion and $110 billion in 1995, all expressed in 1982 dollars (these costs would be even higher in 1990 or 1995 dollars, of course). The foregoing analysis indicates that we will be consistently paying more for oil imports throughout the 1990s than we are currently, and that by 1995 we will be paying even more than we did in 1980–81. It must be noted that the DOE price estimates used in table 3 are based on the assumption that no world oil supply disruptions will occur. Supply restrictions could significantly increase world oil prices and oil import payments.

The most direct cost of imported oil is simply the export of U.S. dollars to pay for the oil. As the price of imported oil increases, larger amounts of American income and wealth must be traded for imported oil. If this money were permanently removed from the U.S. economy, severe trade deficits would be likely. Large trade deficits could increase interest rates, which in turn could lower domestic economic growth and employment. Table 4 lists two commonly used measures of international trade.[28] The merchandise trade balance is the difference between the value of all U.S. goods exports minus the value of all U.S. goods imports. The current account balance includes the differences in value of all U.S. goods *and* services exports and imports, and

TABLE 3. Projected Future Costs of Net Oil Imports to the United States

| Year | Net Imports (MBPD) | World Oil Price ($ per barrel) | Total U.S. Net Import Bill ($ billion) |
|---|---|---|---|
| 1990 | 5 | 26 (low) | 47 |
| | 5 | 40 (high) | 73 |
| | 9 | 26 (low) | 85 |
| | 9 | 40 (high) | 131 |
| 1995 | 5 | 31 (low) | 57 |
| | 5 | 60 (high) | 110 |
| | 9 | 31 (low) | 102 |
| | 9 | 60 (high) | 197 |

TABLE 4. U.S. International Transaction Balances for 1972 to 1984 ($ Billion)

| Year | Merchandise Trade Balance | Current Account Balance |
|---|---|---|
| 1972 | −6.4 | −5.8 |
| 1973 | 0.9 | 7.1 |
| 1974 | −5.5 | 2.0 |
| 1975 | 8.9 | 18.1 |
| 1976 | −9.5 | 4.2 |
| 1977 | −31.1 | −14.5 |
| 1978 | −34.0 | −15.4 |
| 1979 | −27.6 | −1.0 |
| 1980 | −25.5 | 0.4 |
| 1981 | −28.1 | 4.6 |
| 1982 | −36.4 | −11.2 |
| 1983 | −61.1 | −41.6 |
| 1984 | −107.4 | −101.6 |

also includes incoming and outgoing revenue flows due to military sales, technical services, and investment income. As table 4 shows, the United States has had high merchandise trade deficits since 1977. Our annual oil import bills of $40 billion to $80 billion since 1977 have been important contributions to these high merchandise trade deficits. Despite these large trade deficits, however, until recently our current account balance had been positive, except for the period 1977–79. It appears that in the past our large imported oil expenditures have generally been counterbalanced by large increases in U.S. assets abroad, investment income, and the sale of military hardware and technical services.

The trends in our international trade balances have taken an ominous turn since 1982, however. Our merchandise trade balance deficit increased from a previous high of $36.4 billion in 1982 to $61.1 billion in 1983, an increase of 68 percent. In 1984 the merchandise trade deficit soared to $107.4 billion, nearly triple the level of the then-record deficit in 1982. Furthermore, these huge merchandise trade deficits are now causing large current account deficits. In 1982 the United States recorded an overall current account deficit of just over $11 billion. In 1983 our current account deficit rose to $41.6 billion, and in 1984 it totaled an incredible $101.6 billion. Recent trends indicate not only that our relative trade position is plummeting, but also that our positive balances with respect to technical services and investment income are becoming dwarfed by the massive trade deficits. These large merchandise trade and current account deficits siphon off domestic investment capital and force higher interest rates to attract foreign capital to correct the imbalances. The smaller domestic capital pool and higher interest rates reduce domestic investment and employment. In short, large merchandise trade and current account deficits threaten sustained economic recovery.

In addition to the direct economic costs of imported oil, including the export of scores of billions of dollars of U.S. capital and the deleterious impacts on balance of trade, there are several indirect economic costs to American society of imported oil. Evidence strongly suggests that oil imports, particularly in times of rising prices, can have serious secondary macroeconomic impacts on inflation, economic growth, employment, etc. Oil prices significantly increased twice in the last decade, quadrupling in 1974 in response to the OPEC oil embargo and doubling in 1979–80 due to the Iranian oil cutoff. The United States suffered through two periods of double-digit inflation during the last decade, in 1974–75 and in 1979–81, with the beginnings of the inflationary periods coinciding with the oil price shocks. Similarly, the nation has lapsed into its two most severe recessions since the Depression during the last decade, in 1974–76 and 1980–82, as indicated by low or negative real gross national product (GNP) growth and high unemployment. Again, the recessions followed and worsened soon after the higher oil prices and inflation. While there is no absolute proof that higher oil prices necessarily caused these macroeconomic impacts, the evidence is fairly convincing that they were a contributing factor. Other indirect economic costs include simply the impact of high import levels on the price of world oil (higher demand leading to higher prices for all oil imports) and the inevitable price rises which would accompany any sudden supply disruption.

Finally, there are definite national security costs related to our dependence on imported petroleum. Since most of our conventional military vehicles run on petroleum, there is concern about our military readiness in times of supply disruption. Our dependence on Middle Eastern supplies results in direct expenditures for weapons systems, bases, and personnel which are necessary to protect that part of the world, such as the Rapid Deployment Force. Also important are the costs to society of the constraints on our freedom of action in foreign policy due to our dependence on specific Middle Eastern countries.

Economists have developed the concept of "social premium" to account for these costs to U.S. society of relying on imported oil which are not reflected in the direct marketplace price (the total or real cost to society is then the sum of the market price and the social premium). There is no consensus on the overall value of the social premium for imported oil. Estimates have ranged from as low as $5 per barrel to as high as $72 per barrel.[29] Many of these estimates excluded one or more of the indirect costs discussed here. A 1982 Office of Technology Assessment report did not give a specific estimate due to the uncertainties involved, but did state that "the possible future import premium could range up to $50 per barrel."[30]

In summary, it is clear that oil imports represent a major burden on the domestic economy. The United States has paid as much as $79 billion per year to foreign nations to satisfy its petroleum appetite, and projections are that we will exceed this level by the early 1990s even assuming no major supply disruptions. These payments involve a major export of potential investment capital, and threaten to seriously worsen our already perilous international trade situation. There is agreement that there is a social premium associated with our dependence on imported oil, and that the "total" cost of imported oil to the United States is much higher than the market price, possibly by as much as a factor of two or three.

Given that the United States needs to reduce its consumption of petroleum, the most important end-use sector to focus on is transportation. There are two fundamental reasons for this. First, 61 percent of all petroleum consumed in the United States is utilized by transportation.[31] Second, those uses that account for the remaining 39 percent of petroleum consumption typically have energy alternatives to petroleum. The use of distillate fuel oil to heat residential and commercial buildings is declining as users switch to natural gas and electricity (and even wood). Petroleum usage by electric utilities is being phased out in view of the better economics of coal combustion. The petrochemical industry can consider natural gas or coal-based feedstocks. Transportation, on the other hand, currently has no real alternative to petroleum. The percentage of petroleum consumed by transportation will likely increase in the future. Accordingly, any program to significantly reduce petroleum consumption in the United States must necessarily focus on providing an alternative fuel that can be used to power transportation vehicles.

# CHAPTER 2 The Optimum Solution

Chapter 1 summarized two critical problems facing the United States—acid rain and dependence on imported petroleum. A solution for acid rain must involve significant reductions in sulfur dioxide emissions and should minimize cost and market disruption. Reducing our dependence on imported oil will require the development of an alternative fuel that can be economically competitive with petroleum when used in motor vehicles and that can be produced from domestic feedstocks. Though the acid rain problem has been recognized for several years, and energy independence has been a national priority for a decade, little progress has been made by either the private or public sectors in formulating satisfactory long-term solutions for these problems.

There is a fundamental link between the acid rain and imported petroleum problems which facilitates the development of an integrated solution more easily than if either problem were addressed separately. This link is the utilization of the U.S. high-sulfur coal resource base for the production of methanol for use as a transportation fuel.

Coal played an instrumental role in America's industrialization. As early as 1885 coal supplied one-half of U.S. energy needs. Coal's percentage of our total energy supply peaked in 1910 when it accounted for 75 percent of our energy consumption. Although absolute coal production grew between 1910 and 1920, coal's fraction of the energy market actually declined as oil and natural gas became available. Due to their convenience, competitive prices, and newly found reserves, oil and natural gas began to replace coal in both the transportation (railroad) and household sectors. By the 1960s, coal was used nearly exclusively for electrical generation and in industrial applications. In 1982, U.S. coal consumption totaled 707 million tons, of which 84 percent was used in utility power plants and 15 percent was used by industry. Only 1 percent of all coal was used for residential or transportation purposes.[1]

Any overall rationalization of long-term U.S. energy policy must recognize that coal will play an important role because of our huge domestic coal reserves. The Department of Energy has estimated that there are approximately 473 billion tons of coal in the Demonstrated Reserve Base and the

U.S. Geological Survey has estimated that there are another 3,900 billion tons of undiscovered coal resources in the United States.[2] Assuming recovery rates of 60 percent for demonstrated reserves and 30 percent for undiscovered resources, there are 1,450 billion tons of recoverable coal in the United States. This value dwarfs current annual consumption of 700 million tons.

Moreover, the potential energy available from U.S. coal is massive compared to other domestic fossil fuel resources. Table 5 shows the distribution of U.S. recoverable fossil fuel resources among coal, oil shale, crude oil, conventional natural gas, and unconventional gas based on two different measures of economic feasibility for oil shale. Including only those oil shale deposits with over 30 gallons of oil per ton, coal accounts for 91 percent of all recoverable U.S. fossil fuel resources. Even if we use 15 gallons of shale oil per ton as the feasibility limit, coal still accounts for nearly 82 percent of U.S. recoverable fossil fuel resources.[3] The unmistakable conclusion from table 5 is that coal must play a significant role in any long-term U.S. energy policy barring a rapid and comprehensive conversion to either nuclear power or renewable energy sources, neither of which appears likely given current economic and political realities. More to the point, coal is the one domestic energy resource we have in quantities that would permit it to be utilized as a long-term feedstock for our transportation system.

The primary manner in which we do use coal, in electrical generating power plants, is believed to be the primary cause of acid rain. Nearly 70 percent of all sulfur oxide emissions in the eastern half of the United States are from coal-fired electric utilities. A majority of the coal reserves in this country, particularly those in the West and Southeast, contain low levels of sulfur and would result in relatively low levels of sulfur dioxide emissions. Unfortunately, the geopolitical reality is that high-sulfur coal reserves are located in areas very close to large population centers and our industrial heartland. The livelihoods of scores of thousands of families and hundreds of communities in

TABLE 5. Recoverable Fossil Fuel Resource Distribution in the United States

| Resource | Percentage of Total Recoverable Fossil Fuel Energy[a] | Percentage of Total Recoverable Fossil Fuel Energy[b] |
|---|---|---|
| Coal | 91.2 | 81.7 |
| Oil shale | 2.8 | 12.9 |
| Crude oil and natural gas liquids | 2.2 | 2.0 |
| Conventional natural gas | 2.2 | 2.0 |
| Unconventional gas | 1.6 | 1.4 |

[a]Including only those oil shale resources containing over 30 gallons of oil per ton
[b]Including only those oil shale resources containing over 15 gallons of oil per ton

the Midwest and in Appalachia depend upon the maintenance of the high-sulfur coal market.

One view of our energy problem is that we have failed to find a way to utilize our most abundant fossil fuel resource, coal, as a transportation fuel. On the other hand, the combustion of high-sulfur coal in midwestern power plants is the single most important contribution to acid rain. A solution to both of these problems is to burn low-sulfur coal in utility power plants and to convert the displaced high-sulfur coal to methanol for use as a transportation fuel. This strategy would provide the most cost-effective means of reducing sulfur dioxide emissions and acid rain, without decreasing the market for high-sulfur coal, and would provide the nation with a liquid fuel alternative to petroleum that could be produced from our most abundant fossil fuel resource.

The transportation fuel market can easily accommodate large volumes of methanol fuel. Most analyses of major acid rain control programs that would allow least-cost compliance, or fuel switching, have projected that between 90 million and 170 million tons of eastern and midwestern high-sulfur coals would be displaced.[4] It would take 130 million tons of bituminous coal to produce approximately 1.5 MBPD of methanol. Total U.S. highway fuel consumption in 1995 is projected to include 4.5 MBPD of gasoline and 1.6 MBPD of diesel fuel.[5] Assuming that it will take 1.7 times as much methanol to produce the same mileage as gasoline and 2.3 times as much to equal the mileage of diesel fuel (because of the lower volumetric energy content of methanol), 1.5 MBPD of methanol would supply approximately 20 percent of projected 1995 gasoline consumption or just 13 percent of all projected 1995 highway fuel consumption. Once the methanol transition gathered momentum, it can be seen that the potential exists for significant growth in high-sulfur coal demand.

The key to this strategy is the transition to pure methanol as a national motor vehicle fuel. This raises several critical questions: Is large-scale methanol production from coal feasible? Would methanol be economically competitive with gasoline and diesel fuel? What are the relative environmental impacts of methanol compared to petroleum fuels?

The production of methanol from coal involves a two-step process. The first step is coal gasification, whereby the coal undergoes partial combustion with oxygen or air at high temperatures to produce a gas mixture containing primarily carbon monoxide and hydrogen. It is after the coal has been gasified that impurities such as sulfur can be relatively easily and efficiently removed. This ease of sulfur removal permits the production of a sulfur-free fuel from a high-sulfur feedstock. In fact, the sulfur can be recovered and sold for use in the production of sulfuric acid and other chemicals. The second step converts the carbon monoxide and hydrogen mixture to methanol by passing it over a catalyst under pressure. Another appealing aspect of the methanol production process is that the expensive catalysts are poisoned by exposure to sulfur. Thus, there is an inherent internal motivation for the methanol producer to remove all sulfur at the gasification step.

Both steps of the coal-to-methanol production process are considered to be technologically proven. Coal gasifiers are used all around the world in the production of ammonia, and the Sasol-I coal liquefaction plant in South Africa has operated for over twenty-five years with a large gasifier. Gasifiers that would be used in coal-to-methanol plants would be very similar to those in existence today. The methanol synthesis process is also a proven technology, as it is utilized worldwide to produce methanol from natural gas. Thus, the basic technological functions of the coal-to-methanol process are feasible today. This is not the case for other types of coal liquefaction technologies that are often considered to be alternatives to methanol.

The single most critical issue, of course, is the economic competitiveness of methanol with petroleum fuels. Figure 2 shows predicted retail fuel prices, excluding taxes, for methanol from coal, synthetic gasoline from coal, and gasoline from crude oil.[6] The methanol costs range from $7.90 to $15.00 per million Btu in 1982 dollars. The Office of Technology Assessment has projected that methanol could be sold for between $10.20 and $15.00 per million Btu in 1982 dollars.[7] Based on the various studies which have been performed, we project an average value of $12.50 per million Btu for the future retail cost of coal-based methanol. Since there are approximately 56,600 Btu per gallon of methanol, this projected price is equivalent to $0.71

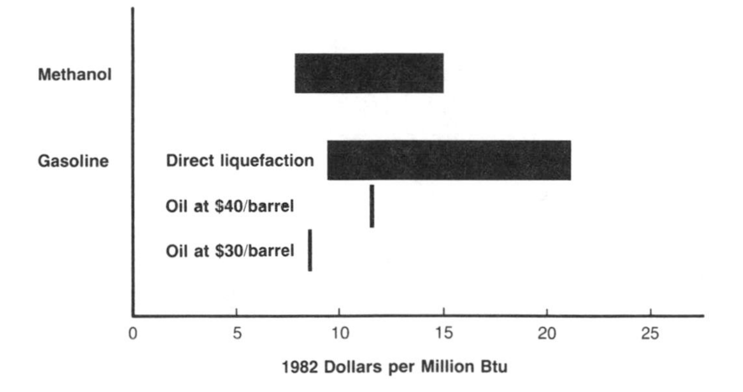


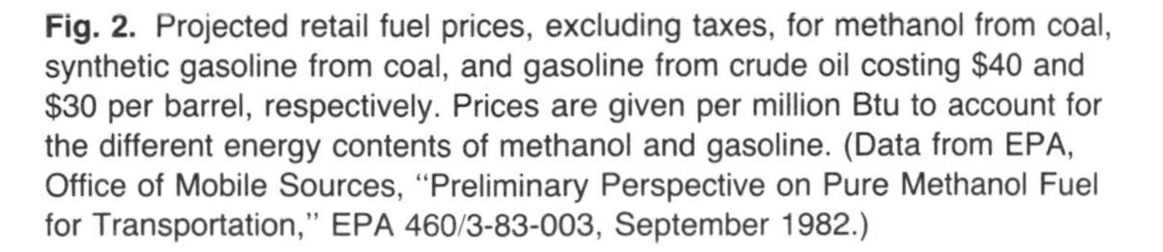
**Fig. 2.** Projected retail fuel prices, excluding taxes, for methanol from coal, synthetic gasoline from coal, and gasoline from crude oil costing $40 and $30 per barrel, respectively. Prices are given per million Btu to account for the different energy contents of methanol and gasoline. (Data from EPA, Office of Mobile Sources, "Preliminary Perspective on Pure Methanol Fuel for Transportation," EPA 460/3-83-003, September 1982.)

per gallon of methanol. Gasoline contains twice as much energy (i.e., twice as many Btu) per gallon as methanol, so we project that methanol would be competitive with gasoline priced at $1.43 per gallon, excluding taxes and assuming that engines utilizing methanol would have the same efficiencies as those using gasoline. Based on historical prices, a gasoline price of $1.43 per gallon, excluding taxes, would be expected when the price of crude oil reaches $43 per barrel.

Methanol is an excellent automotive fuel which yields improved performance and efficiency in Otto-cycle (i.e., gasoline-type) engines. Testing of prototype methanol vehicles has generally shown them to be up to 15 percent more efficient than similar gasoline vehicles on a Btu basis (it is appropriate to compare methanol and gasoline on an energy efficiency basis and not on a fuel economy basis since methanol contains one-half the energy per unit volume of gasoline).[8] But methanol fuel has several unique properties which are not fully exploited by current methanol engines and which offer the opportunity for efficiency improvements. For example, it is well known that engine efficiency increases at leaner air-fuel ratios. Gasoline cannot be operated very lean because of engine misfire. Methanol's higher flame speed allows it to maintain stable combustion at much leaner air-fuel ratios. Preliminary research has supported the contention that methanol can be successfully combusted under very lean air-fuel conditions. In addition, there are other technologies such as turbocharging, supercharging, and methanol fuel dissociation which have only been briefly investigated but which offer additional efficiency improvements. We believe an optimized high-compression, lean-burn Otto-cycle methanol engine will ultimately be 20 to 30 percent more efficient than its gasoline counterpart. Thus, from the viewpoint of the overall cost to the motorist, methanol would be more economical than gasoline from $43-per-barrel oil, and would be competitive with gasoline from oil at approximately $35 per barrel. Referring again to table 3 we can see that world oil prices are expected to rise above this level in the 1990s even assuming no world oil supply disruptions. Thus, it appears that methanol will be very competitive with gasoline in the 1990s, and sooner if world oil supply constraints develop. Given the long lead times associated with a transition to a new transportation fuel, major efforts must be undertaken very soon if a significant supply of fuel methanol is to be available by the mid-1990s.

Methanol has always been considered a very clean-burning fuel; in fact, its low emissions (especially of nitrogen oxides) was a primary impetus for many of the initial methanol engine and vehicle research projects in the early 1970s. Compared to gasoline, methanol engines emit significantly lower levels of nitrogen oxides, though the use of reduction catalysts on gasoline vehicles can result in similar levels of tailpipe emissions (one potential economic advantage of methanol is that catalytic converters may not need to be as complex as with gasoline vehicles, resulting in savings in the cost of the emission control system). The most important air quality benefit of methanol substitution for gasoline may be lower ozone levels. Gasoline vehicles emit a large number of hydrocarbon compounds which are very reactive in the photochemical oxidant (smog) process. Fuel-related emissions from methanol vehicles are comprised almost exclusively of just two compounds, methanol and formaldehyde. Because approximately 90 percent of methanol fuel–related emissions is unburned methanol, which has a low photochemical reactivity, while only the remaining 10 percent, formaldehyde, is highly reactive, it would be expected that the overall reactivity of methanol exhaust would be lower than that of gasoline exhaust. Initial computer modeling studies have projected peak ozone reductions of 15 to 25 percent in Los Angeles if all motor vehicles were methanol-fueled.[9] If further modeling and smog chamber tests confirm these findings, this would be a significant benefit of methanol usage since ozone continues to be a persistent air quality problem in nearly all of our largest metropolitan areas.[10]

The environmental benefits of methanol are most evident when methanol is considered as a substitute for diesel fuel, which is the dominant fuel for large trucks and buses and is becoming more popular for smaller trucks. Diesel engines produce high levels of particulate and nitrogen oxide emissions, and trucks and buses are major sources of these pollutants in many urban areas. Historically methanol has been considered a poor diesel engine fuel because of its very low cetane number. In the past few years, however, engine researchers have found several ways to overcome methanol's autoignition difficulties, and methanol is now considered to be a fine diesel engine fuel. Prototype diesel-cycle methanol-fueled vehicles have achieved energy efficiencies similar to those of diesel-fueled vehicles, and it is possible that optimized methanol vehicles may be even more efficient. Methanol engines emit very low levels of the two most important pollutants from diesel engines, particulates and nitrogen oxides. Preliminary testing has indicated that methanol engines may be able to reduce particulate and nitrogen oxide emissions by 90 percent and 50 percent, respectively, compared to uncontrolled diesel engines.[11] Methanol substitution for diesel fuel would also reduce reactive hydrocarbon emissions (and thus ozone) and sulfur dioxide emissions (diesel fuel contains a small percentage of sulfur while methanol does not). The use of methanol in large trucks and buses could result in a significant improvement in urban air quality.

The use of methanol in motor vehicles would result in increased emissions of methanol and formaldehyde, both of which can be toxic at elevated concentrations, compared to current gasoline vehicles, although methanol vehicles with catalytic converters emit formaldehyde at the same rate as current diesel vehicles and older noncatalyst gasoline vehicles. The determination of the possible public health impacts of higher ambient levels of methanol and formaldehyde has been a prime area of research. Preliminary calculations, utilizing very conservative assumptions, indicate that future projected levels could only approach ranges of concern in certain extreme worst-case exposure situations.[12]

While methanol vehicle emission results to date have been impressive, it is expected that the automotive industry will make substantial improvements. Emissions of gasoline and diesel engines have been research priorities for fifteen years and as a result such engines are highly optimized. Current methanol engines simply involve relatively minor modifications of engines which had been originally designed for gasoline or diesel fuels. Today's methanol vehicles utilize catalytic converters designed for gasoline vehicles; there has been no optimization of emission control systems for methanol vehicles. As the consensus grows that methanol is the transportation fuel of the future, it would be expected that the automotive manufacturers would invest greater resources in methanol engine and emission control system designs, which should lower emissions and increase fuel economy. Based on the low emission levels of current methanol vehicle prototypes and the high likelihood of further progress in methanol engine and emission control system designs, it is clear that the use of methanol as a motor vehicle fuel would be very beneficial for urban air quality.

In summary, we believe there is a very promising solution to both the acid rain and imported oil problems that simply involves a rational utilization of our national coal resources. Using our massive low-sulfur coal reserves to fuel power plants in the Midwest is clearly the most efficient way to reduce acid deposition. We can maintain (and ultimately increase) the market for high-sulfur coal by utilizing it as a methanol feedstock, providing our country with a long-term liquid fuel alternative that would allow us to wean ourselves from our economic and military dependence on imported oil. The production of methanol from coal is technologically feasible today, and will be economically competitive with gasoline by the time large coal-to-methanol plants could be built, even assuming no disruption of world oil markets. The utilization of methanol as a transportation fuel could provide significant vehicle efficiency improvements and improved air quality in urban areas.

# CHAPTER 3 The Methanol Implementation Program

The conclusion that the rational use of domestic coal resources could solve both our imported petroleum and acid rain problems is straightforward. The key is the transition to methanol as a motor vehicle fuel. There is increasing agreement that methanol is the most promising future transportation fuel. Ford Motor Company has stated in congressional testimony that

> we believe that near-neat methanol has the best long-range potential for transportation applications . . . and has the potential to deliver more miles per unit of resource at a lower cost per mile than gasoline or synthetic gasoline.[1]

General Motors' director of energy economics has said "we could get as much as 30 percent more mileage per ton of coal from coal-methanol systems than a coal to gasoline system."[2] Many energy industry representatives have expressed similar views.

While the private sector has recognized the promise of methanol, there is considerable inertia in the existing petroleum-based transportation system. It would be exceedingly difficult for any alternative fuel to break into the transportation market if it were not already conveniently available to the consumer. Major private investments are necessary in several areas—fuel production, fuel distribution, vehicle supply—and the timing would need to be such that all these investments would come together simultaneously in order for the process to work. It is a classic "chicken-and-egg" dilemma. Fuel suppliers are understandably reluctant to invest billions of dollars in coal-to-methanol plants without certainty that vehicles will be able to use the fuel, and automotive manufacturers are not going to commit to methanol vehicle production until they are assured that the fuel will be conveniently available to consumers. Moreover, unless fuel and vehicle manufacturers are very confident that methanol fuel and methanol vehicles will be simultaneously available to the public the investment process will be paralyzed.

Some have suggested that the economic, energy, and environmental benefits of methanol's use as an automotive fuel are so compelling that ultimately the marketplace will dictate its use. Such reasoning continues that all government needs to do is remove any regulatory or institutional barriers that might inhibit the transition to methanol. That certain regulations may affect methanol's use as a fuel, or that the lack of regulations in certain areas might lead to uncertainties with respect to future regulatory treatment, cannot be denied. In 1983 the General Accounting Office completed a helpful study on issues such as the establishment of Environmental Protection Agency (EPA) emission certification test procedures and standards, the determination of an appropriate methodology for inclusion of methanol vehicles in the corporate average fuel economy program, and the equitable taxation of methanol fuel.[3] EPA has recently initiated a process to establish methanol vehicle certification test procedures and standards as well as a methodology to include methanol vehicles in the corporate average fuel economy program.[4] The critical point, however, is that even if all regulatory uncertainties were removed, there would still be the critical implementation questions concerning the timing and magnitude of private sector investments. These are the true implementation barriers to the methanol transition, and many who have studied this issue have concluded that some type of governmental coordination will be required for an efficient and nondisruptive transition. For example, Donald Petersen, chairman of Ford Motor Company, has stated that "America won't make the change to methanol without sensible government initiatives to make it more attractive than continued consumption of petroleum-based fuels."[5]

Because of the chicken-and-egg dilemma, as well as the critical importance of the timing of various investment decisions, we believe that the following incentives are necessary to encourage both the production and consumption of methanol as a transportation fuel. The incentives to produce methanol are: (1) a highway fuel tax, (2) a methanol price guarantee program, and (3) a methanol availability requirement at certain service stations. The incentives to promote methanol fuel demand are: (1) a federal tax credit for new methanol vehicle purchases, (2) a requirement that all new urban transit buses utilize methanol fuel, and (3) a requirement that all new federal fleet vehicles be methanol-fueled.

The following sections introduce each of these incentives. A quantitative analysis of the economic impacts of the integrated program will be presented in chapter 4. For more detailed descriptions and analyses of each of the individual incentives, the reader should consult the appropriate chapters in Part 2.

## Highway Fuel Tax

Some form of petroleum tax is necessary to stimulate the production of methanol from U.S. coal. The uncertainty over future petroleum prices is probably the greatest barrier to the methanol transition. While methanol from coal will be economically competitive with gasoline when crude oil is approximately $35 per barrel, and nearly all experts predict world oil prices to rise above this level within a decade at most, no group of investors wants to be the first to take the risk of spending billions of dollars on a coal-to-methanol facility faced with uncertain future petroleum prices. Investors also fear that petroleum-exporting countries might lower their oil prices as these new coal-to-methanol plants begin producing methanol, thus undercutting the methanol fuel market.

There are, of course, reasons to reduce the level of oil imports, as a petroleum tax would do, aside from its relationship to the methanol implementation program. The cost of imported oil is enormous. As shown in chapter 1, the export of U.S. dollars for petroleum is expected, even without possible supply disruptions, to reach $100 billion annually by the early 1990s. The huge payments for imported oil represent the key element in our balance of trade problem. The effects of importing such quantities of petroleum are manifest in inflationary pressures, reduced economic growth, and unemployment. Many have estimated the social cost of imported petroleum to be an additional $10 to $50 per barrel over the actual market price.

Faced with massive federal deficits, a petroleum tax should be considered as an important alternative to increasing personal income taxes. Taxes on consumption, such as a petroleum tax, are preferred by the public over non-discretionary taxes such as income taxes.

A petroleum tax could take one of two basic forms. The tax could be levied on imported oil or the tax could be more directly applied to a consuming sector, such as a highway fuel tax. Purely on the technical merits, an oil import tax deserves serious consideration. But based on the strong political opposition that would be mounted against such a tax from U.S. petrochemical and other manufacturers whose products utilize a significant amount of petroleum, from those in the Northeast who use fuel oil for heating, and from some oil companies that import most of the petroleum they market, some form of highway fuel tax appears more realistic. While a highway fuel tax would not encourage efficient use of all petroleum products, it would provide a stimulus to develop methanol as a transportation fuel. Since this is the primary objective in the design of the methanol implementation program, a highway fuel tax is specifically proposed.

It is possible to eliminate oil imports entirely by a very large highway fuel tax and concurrently raise massive federal revenues, but such a tax would have serious impacts on standards of living, inflation, and the U.S. economy as a whole. While several options have been considered, our recommendation is for a rather modest tax. The philosophy behind the highway fuel tax would not be to eliminate petroleum use in transportation per se, but rather it would be to cap or establish a ceiling for the price of oil imported into the United States. The specific proposal is to phase the tax in over a seven-year period with the tax reaching a maximum in the seventh year of $0.40 per gallon. This tax would cap oil import prices at $30 per barrel, since with the tax, petroleum costing $30 per barrel would yield gasoline with a retail price equivalent on an energy basis to the projected retail price of methanol from coal. If oil producers chose to increase the price of oil above $30 per barrel this would stimulate large and irreversible investment in coal-to-methanol plants which would further reduce their market.

Perhaps the most important design feature of the highway fuel tax proposal is that it would be phased in over a seven-year period. The phase-in feature minimizes the inflationary and other impacts of such a tax by providing time for highway fuel users to purchase more fuel-efficient vehicles. Over a seven-year period, vehicle manufacturers and consumers alike can make the transition to fuel-efficient and methanol-fueled vehicles in an orderly and efficient manner.

### Methanol Price Guarantee

There still exists the possibility, especially with only a modest highway fuel tax, that petroleum-exporting countries would undercut the price of methanol from coal during that critical period when the initial methanol plants were coming on line. Although such an outcome would save billions of dollars for petroleum consumers and the country as a whole, it would be devastating to the coal-to-methanol plant investors. Also, with the remaining uncertainties regarding the development of the demand for methanol fuel (e.g., when will large numbers of methanol vehicles be produced?), a coordinated plan for the supply and demand for methanol is necessary.

Although there are a number of options available, a temporary, competitive, fixed-price guarantee for methanol fuel production appears to be the best way to minimize the risk for investors and control the implementation timing, while at the same time minimize government involvement and costs and stimulate the market to work in the most efficient way possible. For the program we recommend, the federal government would invite bids for a price guarantee of $0.50 per gallon of fuel grade methanol delivered to St. Louis, Pittsburgh, or Atlanta. The government would accept bids at the $0.50 per gallon price up to a maximum new production capacity for that year on a first-come basis. If there were an insufficient number of bids at $0.50 per gallon to reach the capacity ceiling, then the government would issue a second invitation with a guarantee of $0.55 per gallon. Successive invitations at price increments of $0.05 per gallon would be issued until the capacity ceiling was reached, although no invitation would be issued with a guarantee in excess of $0.80 per gallon. The final guaranteed price for all projects for that year would be equal to the guaranteed level in the final successful bid. Contracts would be signed between the government and the approved project sponsors for the agreed price level, methanol volume, and time frame. When the contract becomes due, the government's involvement is enforceable only if the project is producing methanol fuel. The producer sells the methanol in the marketplace. If the market price is higher than the guaranteed price, then there is no governmental liability whatsoever. If the market price is less than the price guarantee, then the government must reimburse the producer for the difference for the quantity of fuel specified in the agreement. The price guarantee would apply for five years after the contractually agreed production start-up date and only for methanol produced from eastern and midwestern coals with sulfur contents greater than 1 percent.

To coordinate the implementation timing and permit an orderly phase-in, the methanol price guarantee should cover 300,000 barrels per day (BPD) of new methanol production capacity each year for five successive years. This annual increase in capacity would be sufficient to fuel about 5 million new passenger vehicles each year (about one-half the annual production of passenger cars and light trucks). Over the five-year period of awarding price guarantees for new production capacity, a total of 1.5 million barrels per day capacity would be in place. This production capacity would maintain the existing market for midwestern and eastern high-sulfur coal, but would still represent only about 15 percent of total transportation energy needs. The potential exists for a continued expansion of the coal-to-methanol fuel market.

Although a methanol price guarantee is needed to reduce investment uncertainty, actual government price support payments are not likely. The only risk to the government is that oil producers will keep their prices below $30 per barrel which could then necessitate price support payments for the methanol program. If, for example, oil producers offered oil for $25 per barrel (instead of $30 per barrel), the maximum annual price support payment would be about $3 billion, yet the savings to the United States would be over $27 billion annually due to the lower oil price applying to all petroleum consumed.

### Methanol Requirement for Service Stations

Methanol fuel must be conveniently available at retail outlets at the time methanol-fueled vehicles are offered for sale to the public. Although the marketplace would ultimately respond to the need for methanol retailing, precise timing is necessary to coincide with the production of large quantities of methanol fuel and the introduction of large numbers of methanol vehicles. Several strategies are available to the government to try to ensure that service stations offer methanol fuel, but only a federal mandate to require methanol fuel at certain retail outlets achieves the desired goal with maximum certainty and minimum cost. The precedent of the 1973–74 EPA regulations which effectively integrated unleaded gasoline into the national service station network is evidence that a similar requirement for methanol could be successful.

The main feature of this initiative that minimizes its adverse effects is that only the largest stations would be required to offer methanol. Even if only one-half of all service stations were required to offer methanol fuel, methanol would be conveniently available. In addition, the largest service stations would experience less of an economic impact and would likely have greater flexibility with existing facilities to offer methanol with minimal additional cost. It is likely that once the implementation program accelerated even the smaller retail outlets would add methanol fueling capability.

## Methanol Vehicle Tax Credit

The most important element in the demand for methanol is the individual vehicle purchaser. If consumers want methanol-fueled vehicles, the automobile manufacturers will provide them. But while the entire nation will benefit from methanol usage in the long term, most individual vehicle purchasers may not seriously consider buying a methanol vehicle in the initial years of the program. Many people would rather not be the first to try a new product. Others will be concerned about the availability of methanol fuel and about its future price.

An encouragement for the purchase of the initial methanol vehicles will likely be necessary. A federal tax credit for methanol vehicle purchases seems most appropriate. Tax credits are clearly targeted to reduce the vehicle purchase price to the consumer, can be easily and quickly phased out, are easily understood by the public, and can be administered and enforced by the Internal Revenue Service as are other tax credits.

The tax credit should be limited to 5 million vehicles per year to correspond to the projected methanol supply. The tax credit could be made available for only one year or for several years. One option would involve a $500 per vehicle tax credit for methanol vehicles purchased in the initial year up to the maximum of 5 million vehicles. This limited one-year program might be sufficient to ensure significant consumer interest in methanol vehicles if other parameters, such as methanol vehicle sticker prices, world oil prices, and methanol fuel costs, are favorable. These variables, which cannot be accurately projected at this time, may not be sufficient to ensure methanol vehicle demand and a more comprehensive tax credit program would probably be necessary. The specific option being suggested would apply for three years with the first-year tax credit being $1,000 per vehicle, the second-year tax credit being $800 per vehicle, and the final-year tax credit being $600 per vehicle. Providing a high degree of certainty in the purchase of the appropriate number of methanol vehicles during the critical, initial years of the transition is absolutely necessary for its success. It is necessary to assure demand for methanol fuel to avoid the necessity of large government price supports and to assure a known market for new methanol vehicles so vehicle manufacturers can plan for that level of production with confidence. An interesting additional feature of this tax credit is that, since it is using federal tax revenue to stimulate a desired domestic program, the tax credit might be limited only to domestically produced methanol-fueled vehicles.

### Urban Bus Requirement

Urban transit buses are significant sources of particulate matter and nitrogen oxides air pollution in many urban areas. Two other characteristics of diesel combustion, odor and black smoke, are also major irritants to urban dwellers. Because buses operate on busy urban streets and emit their pollutants at ground level, public exposure to bus pollution is relatively high. Methanol combustion produces little or no particulate matter or smoke, and nitrogen oxide levels are approximately one-half those of diesel engines.

The utilization of methanol in urban buses before its widespread availability for passenger cars would be a logical phase-in program for methanol fuel. Methanol-fueled urban buses would be highly visible to the public and would therefore maximize public awareness of methanol as a desirable transportation fuel. There are a number of reasons why a methanol urban bus program could be implemented fairly easily, for example, urban buses are supported largely by government funds and are centrally fueled and maintained. Centralized fueling sites simplify the fuel distribution problem substantially.

We recommend that all urban transit buses purchased in part with federal funds be methanol-fueled beginning three years before the initiation of the large-scale methanol fuel transition program. In addition to the 80 percent share of capital expenses already provided by the federal government, the federal government should also provide 100 percent of the incremental cost of the methanol bus over the cost of a similar diesel bus. The federal government should also reimburse transit authorities for any incremental operating expenses due to the differential cost of methanol and diesel fuel.

### Federal Fleet Requirement

Requiring that new federal fleet vehicle purchases be methanol-fueled could be an easy and important symbolic first step in an overall methanol implementation program. In terms of direct fuel and vehicle demands, the federal fleet can play only a small role, but one that will emphasize that the federal government is serious about the methanol implementation program. Widespread use of federal methanol vehicles, clearly identified, should also reduce consumer apprehension about such vehicles when they are offered for public sale three years later.

## Timing of the Program

Six initiatives have been briefly summarized which should be adopted by the federal government to spur the production and consumption of methanol fuel. The coordination of the incentives with respect to timing is of critical importance. Figure 3 presents an aggressive schedule for the implementation of the initiatives described here. In considering the schedule, the urgency of our situation should be stressed. Considering the lead time necessary to construct methanol plants and begin the methanol fuel transition, even with this most aggressive schedule only 15 percent of our transportation fuel needs would be supplied by methanol by 1994. Therefore, for the United States to have a real alternative to our dependence on imported petroleum within the next ten years, such an aggressive program must begin now.

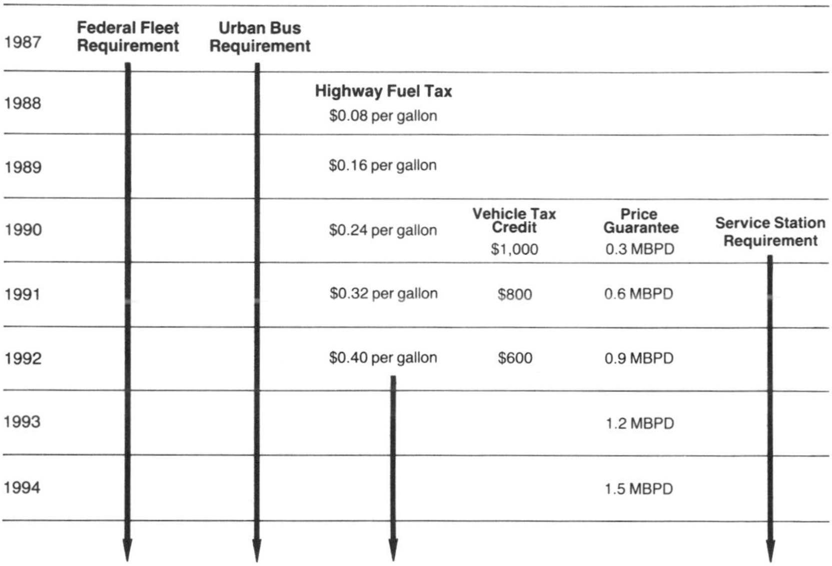

| Year | | | | | | |
|---|---|---|---|---|---|---|
| 1987 | **Federal Fleet Requirement** | **Urban Bus Requirement** | | | | |
| 1988 | | | **Highway Fuel Tax**<br>$0.08 per gallon | | | |
| 1989 | | | $0.16 per gallon | | | |
| 1990 | | | $0.24 per gallon | **Vehicle Tax Credit**<br>$1,000 | **Price Guarantee**<br>0.3 MBPD | **Service Station Requirement** |
| 1991 | | | $0.32 per gallon | $800 | 0.6 MBPD | |
| 1992 | | | $0.40 per gallon | $600 | 0.9 MBPD | |
| 1993 | | | | | 1.2 MBPD | |
| 1994 | | | | | 1.5 MBPD | |

**Fig. 3.** Schedule for implementation of methanol initiatives. All of the initiatives would continue past 1994 except for the tax credit, which would end after 1992, and the price guarantee, which would begin to phase out in 1995 and would end after 1998.

**Overall Methanol Supply and Demand**

This methanol implementation program has been designed to coordinate, to as great a degree as possible, overall methanol fuel supply and demand during the critical transition years of the late 1980s and early 1990s. Three of the implementation initiatives—the federal tax credit for methanol vehicle purchases, the urban bus requirement, and the federal fleet requirement—would directly stimulate demand for methanol fuel. Table 6 projects the annual fuel demands for each of these initiatives. The projected fuel demands for the methanol vehicle purchase tax credit program were based on assumptions that 5 million methanol vehicles would be sold each year from 1990 through 1994, and that the average new methanol vehicle is driven 12,000 miles per year and achieves 15 miles per gallon of methanol (assuming that methanol vehicles are 20 percent more energy-efficient than new gasoline vehicles having an average on-road fuel economy of 26 miles per gallon). It was assumed that methanol buses would have energy efficiencies similar to diesel buses (i.e., methanol bus volumetric fuel consumption would be 2.28 times greater) and that 10 percent of total urban bus fuel consumption would be replaced by new methanol buses each year beginning in 1987 (the urban bus fleet would not be completely methanol-fueled until 1996, when fuel demand would reach 72,000 BPD). The methanol fuel demands for federal fleet vehicles were calculated based on assumptions of a 15 percent turnover of the federal fleet fuel consumption each year beginning in 1987 and methanol vehicles being 20 percent more energy-efficient.

The remaining three initiatives—the methanol price guarantee program, the highway fuel tax, and the methanol availability requirement for high-volume service stations—have been designed to ensure appropriate methanol fuel supply. The price guarantee program would contract for 300,000 BPD of methanol in 1990, 600,000 BPD in 1991, 900,000 BPD in 1992, 1,200,000 BPD in 1993, and 1,500,000 BPD in 1994. It can be seen that these quantities are very similar to the projected methanol fuel demands in table 6 for the years 1990 through 1994. The methanol fuel demands for the years prior to 1990

TABLE 6. Projected Methanol Fuel Demand from 1987 through 1994 for the Methanol Implementation Program (Thousands BPD)

| Initiative | 1987 | 1988 | 1989 | 1990 | 1991 | 1992 | 1993 | 1994 |
|---|---|---|---|---|---|---|---|---|
| Urban bus requirement | 7 | 14 | 22 | 29 | 36 | 43 | 50 | 58 |
| Federal fleet requirement | 5 | 9 | 14 | 18 | 23 | 28 | 31 | 31 |
| Private passenger vehicles (tax credit) | 0 | 0 | 0 | 260 | 520 | 780 | 1,040 | 1,300 |
| Total | 12 | 23 | 36 | 307 | 579 | 851 | 1,121 | 1,389 |

can easily be met by the large current and projected methanol surpluses from natural gas feedstocks. The slight surpluses in the 1991 to 1994 time frame would allow the fuel demands to be met even if the actual outputs of some of the initial coal-to-methanol plants were slightly less than projected, or the surpluses could be used to satisfy other methanol fuel demands not directly associated with the implementation program (such as large trucks or intercity buses). The highway fuel tax does not directly supply methanol, but is necessary to assure investments in large coal-to-methanol plants. Finally, the methanol availability requirement for large service stations provides the necessary distribution link between fuel producer and vehicle operator.

# CHAPTER 4 Overall Effects of the Methanol Implementation Program

We have argued that utilizing low-sulfur coal in electric power plants and converting high-sulfur coal to methanol for use as a transportation fuel would solve both our acid rain and petroleum import problems in an optimum manner. Chapter 3 briefly described six separate initiatives we believe would facilitate and coordinate the methanol transition. This chapter will present the results of our analyses of the impacts of these implementation initiatives on employment, oil import payments and the trade deficit, the federal budget, gross national product, and consumer prices and inflation. It will also briefly summarize the effects of the program on the coal industry, the automobile industry, and air quality.

Three of the initiatives—the service station requirement, the urban bus program, and the federal fleet requirement—are fairly narrow in scope and would not have far-reaching economic impacts. Details of their impacts are presented in Part 2. The remaining three proposals are much broader programs which could have significant macroeconomic impacts. Various options will be discussed in Part 2 for these initiatives, but we will assume the following specific options for the subsequent analyses: a phased-in highway fuel tax which begins in 1988 at $0.08 per gallon and rises to $0.40 per gallon by 1992, with exemptions for commercial trucks with gross vehicle weight ratings in excess of 50,000 pounds; a methanol price guarantee program which will contract for 300,000 BPD of high-sulfur coal-to-methanol production beginning in each of the years from 1990 through 1994 and lasting for five years; and a methanol vehicle purchase tax credit which would be $1,000 in 1990, $800 in 1991, and $600 in 1992 for the first 5 million methanol vehicles sold in each of those years. The anticipated result of implementing all of these incentives would be sufficient private investment to ensure the timely production and consumption of 1.5 MBPD of methanol for use as a transportation fuel by 1994. All of the following analyses were performed in 1982 dollars.

## Employment

The development of a 1.5 MBPD domestic high-sulfur coal-to-methanol fuel production industry would be expected to have a significant impact on employment levels, especially in those regions of the country with large high-sulfur coal reserves. We have worked with Jack Faucett Associates, Inc., an economics consulting firm, to perform an analysis that utilizes economic models to project the total employment which would be associated with a 2.5 MBPD methanol fuel production industry.[1] That analysis assumed that six 85,000 BPD coal-to-methanol plants would come on line each year from 1990 through 1994 for a total of 2.5 MBPD. Employment impacts were calculated beginning in 1985, when construction of the initial plants would begin, and were continued through 1995, the second year for which all thirty plants would be operational. However, our proposed methanol price guarantee program involves a capacity ceiling of only 1.5 MBPD. Accordingly, the employment levels associated with a 1.5 MBPD price guarantee program have been derived and are presented in table 7.

Table 7 details the projected employment impacts in terms of general type of employment (construction, operation, and coal-related), as well as by direct, indirect, and induced impacts. Direct employment is that attributed specifically to either the construction or the operation of the methanol plants (''mine mouth'' coal employment can also be considered to be direct employment since the methanol plant would likely have a long-term contract for a guaranteed coal supply). Indirect employment is that generated by the purchase of goods and services from different industrial sectors of the economy in support of plant construction, operation, and coal mining. Induced employment in the economy as a whole occurs as changes in output and employment result in changes in personal income and increased consumption.

It can be seen in table 7 that from 1985 through 1990 construction-related jobs are the largest fraction of total employment. Construction-related employment would peak in 1989 at 197,000, with 60,000 workers involved in building plants and 137,000 workers employed in other industries supported by the construction, such as industrial machinery, metal products, steel, trucking, cement, etc. During the early 1990s, as plants become operational (and assuming that further construction was not occurring), construction-related employment would fall while operation and coal employment would increase. In 1995, which represents a steady-state situation for the completed plants, there would be 58,000 operation-related jobs, 81,000 coal-related jobs, and 88,000 induced jobs in areas near the methanol plants, coal mines, and support industries. Of particular interest is that there would be 44,000 coal mine mouth jobs in 1995. The total employment impacts would peak at 400,000 in 1989 and equilibrate at 227,000 in 1995.

TABLE 7. Projected Levels of Annual Employment Associated with a 1.5 MBPD Domestic Methanol Fuel Production Industry (Thousands)

| Type of Employment | 1985 | 1986 | 1987 | 1988 | 1989 | 1990 | 1991 | 1992 | 1993 | 1994 | 1995 |
|---|---|---|---|---|---|---|---|---|---|---|---|
| Construction | 43 | 94 | 126 | 169 | 197 | 154 | 104 | 71 | 29 | 0 | 0 |
| Direct | 13 | 28 | 38 | 51 | 60 | 48 | 33 | 23 | 10 | 0 | 0 |
| Indirect | 30 | 66 | 88 | 118 | 137 | 106 | 71 | 48 | 19 | 0 | 0 |
| Operation | 0 | 2 | 6 | 10 | 28 | 40 | 50 | 58 | 65 | 59 | 58 |
| Direct | 0 | 2 | 6 | 10 | 15 | 21 | 25 | 27 | 28 | 29 | 29 |
| Indirect | 0 | 0 | 0 | 0 | 13 | 19 | 25 | 31 | 37 | 30 | 29 |
| Coal | 0 | 3 | 10 | 20 | 34 | 50 | 65 | 75 | 81 | 83 | 81 |
| Mine mouth | 0 | 2 | 5 | 11 | 18 | 27 | 36 | 41 | 45 | 46 | 44 |
| Indirect | 0 | 1 | 5 | 9 | 16 | 23 | 29 | 34 | 36 | 37 | 37 |
| Induced | 22 | 50 | 73 | 105 | 141 | 136 | 126 | 120 | 106 | 88 | 88 |
| Total | 65 | 149 | 215 | 304 | 400 | 380 | 345 | 324 | 281 | 230 | 227 |

Where would these jobs be located? The coal-mining jobs would be in those states with large high-sulfur coal reserves, such as Ohio, Illinois, Indiana, Kentucky, Pennsylvania, and Virginia. Since it is expected that the methanol plants will be constructed very near to coal mines, the direct construction and operation jobs should also be in or near these areas. A large percentage of the induced employment will be in these same communities. The indirect employment impacts would be in industrial sectors such as industrial machinery, steel, metal products, and instrumentation, which are typically concentrated in the industrial Midwest and Northeast (and California). Thus, it is apparent that most of the employment generated by a domestic methanol production industry would be in the high-sulfur coal regions of the Midwest and Northeast. This is very appropriate since a least-cost acid rain control program will reduce coal-related employment in these areas and because these same areas have had very high unemployment levels in recent years. Of course, as part of the integrated national coal strategy, midwestern electric power plants would switch to low-sulfur coal to reduce sulfur dioxide emissions. Thus, areas with low-sulfur coal reserves in the West and Southeast would also have employment benefits associated with this program.

This discussion has assumed that the only methanol plant construction would be those projects covered by the price guarantee program. If the methanol implementation program were successful, as expected, in launching a national transition to methanol as a transportation fuel, then the domestic employment increases would be much larger. Additional methanol plant construction would be expected both in the East and Midwest, to be close to large cities in these areas, and in the West, to take advantage of the vast coal reserves there. Accordingly, additional construction, operation, and coal-mining jobs would be expected in all coal-producing areas of the country, and induced employment would be realized throughout the nation. Overall, total employment levels for a national methanol fuel production industry could be several times the levels shown in table 7.

It can be commented that the expenditure of large sums of capital will always result in the creation of jobs. We believe, however, that the employment effects of a coal-to-methanol fuel production industry are particularly constructive for four reasons. First, by its nature, investment in methanol plants results in a considerable cycling of that money throughout the economy before it is used for personal consumption, i.e., we believe such an investment would have a larger "jobs multiplier" than equal expenditures for other programs. Second, except for construction-related employment, the jobs created by methanol plant investment would be permanent, or sustaining, jobs. Third, unlike some other spending, methanol plant investment results in a product that is useful to society over the long term thus generating additional economic activity. Finally, and probably most important, the funding for the methanol plants can be seen as capital available for U.S. investment as a result of lower U.S. oil import payments due to methanol displacement of petroleum fuels (i.e., money that would have been spent for imported oil is now available for investment in the United States). The projected magnitude of our lower oil import payments will be examined in the next section.

### Oil Import Payments/Balance of Trade

The methanol implementation program will decrease petroleum consumption in two ways. Beginning in 1988 the highway fuel tax will raise the price of gasoline and some diesel fuel which will promote price-induced conservation. Beginning in 1990, the methanol price guarantee and methanol vehicle tax credit initiatives will result in direct methanol substitution for petroleum fuels (as will the urban bus and federal fleet initiatives, to a lesser degree, beginning in 1987).

Table 8 shows the projected petroleum conservation impacts of the methanol implementation initiatives for the years 1988 through 1994. We have worked with Jack Faucett Associates, Inc., to project the effect of price-induced conservation as part of a broad study of overall macroeconomic impacts of highway fuel taxes.[2] These results are based on a $0.40 per gallon tax phased in during the period 1988 to 1992 and on the following assumptions: gasoline and diesel fuel consumption by nonexempted vehicles would be 6.0 MBPD during 1988 and 1989 but would be somewhat lower after 1990 due to methanol substitution; the prices of gasoline and diesel fuel would be based on oil at $33 per barrel during the 1988 to 1992 time frame (excluding the new tax); refiners would maintain their current petroleum products pricing structure; and automotive manufacturer product planning and consumer purchase decisions would be such that overall new car fuel cost per mile is held constant even with the highway fuel tax. As seen in table 8, the higher prices

TABLE 8. Projected Reductions in Oil Import Payments Due to Methanol Implementation Program

| Year | Price-Induced Highway Fuel Conservation (MBPD) | Petroleum Displaced by Methanol (MBPD) | Total Petroleum Conservation (MBPD) | Reduction in Oil Import Payments ($ billion) |
|---|---|---|---|---|
| 1988 | .38 | .01 | 0.39 | 5.7 |
| 1989 | .60 | .02 | 0.62 | 9.1 |
| 1990 | .77 | .16 | 0.93 | 14 |
| 1991 | .88 | .31 | 1.19 | 17 |
| 1992 | .99 | .46 | 1.45 | 21 |
| 1993 | .96 | .60 | 1.56 | 23 |
| 1994 | .94 | .75 | 1.69 | 25 |

of gasoline and diesel fuel would promote conservation of 0.38 MBPD of oil in 1988, rising progressively to 0.99 MBPD by 1992, and then dropping slightly to 0.94 MBPD in 1994 (due to lower baseline petroleum consumption because of increased methanol substitution).

The widespread use of methanol as a motor vehicle fuel beginning in 1990 will result in the direct displacement of petroleum fuels. Our analysis assumes that the highway transportation sector will utilize methanol at the volumes given in table 6 (0.3 MBPD in 1990 rising to 1.4 MBPD in 1994) and that on average it will take 1.7 and 2.3 times as much methanol to travel the same distance as a volume of gasoline and diesel fuel, respectively. Table 8 shows that this results in the displacement of 0.16 MBPD of petroleum in 1990, rising to 0.75 MBPD in 1994.

Total petroleum conservation due to the methanol implementation program is simply the sum of the price-induced conservation and direct petroleum displacement. These levels would equal 0.39 MBPD in 1988, increasing each year to 1.69 MBPD by 1994. Valuing this reduced gasoline and diesel fuel consumption at $40 per barrel in 1982 dollars (U.S. oil prices would be expected to be capped at $30 per barrel but gasoline and diesel fuel are among the most valuable crude components), oil import savings would range from $5.7 billion in 1988 to $25 billion in 1994 as shown in table 8. These savings could be considerably larger if world oil supply constraints were to develop resulting in higher world oil prices.

These large reductions in oil import payments could significantly improve U.S. international trade balances. As discussed earlier, the 1984 U.S. merchandise trade balance deficit was a record $107 billion, and the current account deficit was $102 billion in 1984. The magnitudes of these deficits dwarf values for previous years. While temporary deficits are considered to be tolerable, the long-range picture is not promising. Continued high deficits would result in a shortage of U.S. investment capital, which would require higher interest rates to attract foreign capital. The opportunity to reduce the outflow of American dollars on the order of $20 billion to $30 billion per year by the 1990s, while building an infrastructure which can provide a long-term alternative to imported petroleum, is an important benefit of the methanol implementation program.

### Federal Budget

Two of the methanol implementation initiatives—the highway fuel tax and the methanol vehicle tax credit—would have major effects on the federal budget. The highway fuel tax would raise governmental revenue while the methanol vehicle tax credit would lower revenue; the magnitudes of these impacts are fairly easy to predict with a high degree of confidence. The methanol price guarantee program is not expected to result in governmental expenditures, given the existence of the highway fuel tax and projected world oil prices. Two of the remaining initiatives—the urban bus and federal fleet requirements—would each require annual governmental expenditures on the order of $100 million or less. The methanol requirement for service stations would require no federal expenditures. Because of the relative revenue impacts, the following analysis will focus on the highway fuel tax and methanol vehicle tax credit programs.

Table 9 summarizes the major federal budget impacts of the methanol implementation program. The net federal revenue impacts of the highway fuel tax were based on a phased-in $0.40 per gallon tax.[3] The analysis included not only the direct revenues from the tax itself, but also the indirect revenue losses due to highway fuel conservation (i.e., the federal government would lose $0.09 per gallon, the current federal fuel excise tax, on fuel which is conserved) and slightly lower overall tax collections. The results also account for the petroleum which is displaced by methanol after 1990. Still, it can be seen that the highway fuel tax generates large revenues for the federal government. Net revenues peak at $24.8 billion when the tax reaches $0.40 per gallon in 1992, and then fall slightly thereafter as petroleum continues to be displaced by methanol.

The methanol vehicle purchase tax credit would decrease the amount of tax revenues collected by the federal government. The calculation of these tax expenditures is straightforward. Assuming that 5 million methanol passenger vehicles are sold in 1990, 1991, and 1992, with tax credits of $1,000, $800, and $600 per vehicle in these three years, respectively, then the total federal revenue losses would be $5 billion, $4 billion, and $3 billion.

As a part of the methanol price guarantee program described in Part 2, we recommend that the current exemption from federal fuel excise taxes for pure alcohol fuels produced from feedstocks other than petroleum and natural gas be extended from 1988 through 1993 for methanol produced from high-

TABLE 9. Projected Total Federal Budget Impacts of the Methanol Implementation Program ($ Billion)

| Year | Highway Fuel Tax | Methanol Vehicle Tax Credit | Methanol from High-Sulfur Coal Fuel Tax Exemption | Total |
|---|---|---|---|---|
| 1988 | 6.3 | 0 | 0 | 6.3 |
| 1989 | 11.9 | 0 | 0 | 11.9 |
| 1990 | 16.7 | −5 | −0.2 | 11.5 |
| 1991 | 21.1 | −4 | −0.4 | 16.7 |
| 1992 | 24.8 | −3 | −0.6 | 21.2 |
| 1993 | 24.2 | 0 | −0.8 | 23.4 |
| 1994 | 23.5 | 0 | 0 | 23.5 |

sulfur coal. Assuming that the methanol fuel demands given in table 6 will be met entirely by methanol from high-sulfur coal, in accordance with the methanol price guarantee program, and that in absence of the exemption methanol would be taxed on an energy equivalent basis to gasoline, table 9 shows federal revenues would be reduced by $0.2 billion in 1990 (when the initial coal-to-methanol plants begin production), increasing to $0.8 billion in 1993 (the final year the exemption would apply).

Table 9 shows that the total federal revenue effects are both positive and large. Net federal revenues would increase over $6 billion in 1988, and would nearly double to $12 billion the next year. The increased revenues would peak at $23.5 billion in 1994 when the only impact would be the highway fuel tax revenues. The increase in federal revenues due to the tax would drop slightly after 1994 due to slightly lower petroleum consumption.

The increased federal revenues shown in table 9 would be a major benefit of the methanol implementation program. The fiscal year 1984 federal budget deficit was approximately $200 billion and current projections are that deficits will remain in the $150 billion to $200 billion range for at least several years. Some experts believe the deficit will be even higher during the late 1980s. Many economists consider high deficits to be a primary cause of high interest rates and believe that there is a low likelihood of a sustained economic recovery as long as federal deficits remain high. The highway fuel tax will offset a significant part of our overall federal budget deficit, and will be an important step in showing the sincerity of our political leaders in resolving the deficit problem. Although increased taxes are never popular, we believe the imposition of a highway fuel tax is politically feasible for two reasons. First, it would be seen as an integral part of the overall methanol implementation program to stimulate domestic energy production and employment and decrease the export of American dollars. Second, the U.S. public has traditionally preferred taxes on consumption to other types of levies (such as income or property taxes). Consumers would have the ability to ameliorate the impacts of the phased-in tax by switching to a higher fuel economy or methanol-fueled vehicle.

One effect that has not been taken into consideration in the analyses which led to table 9 was the beneficial impact on the federal treasury of increased domestic investment and employment stimulated by the development of a domestic methanol fuel production industry. Increased employment would not only broaden the federal tax base, but would also reduce federal expenditures for unemployment insurance and other social welfare programs. The precise magnitudes of these impacts are impossible to project, but could be very substantial especially if the initial program launches a nationwide transition to methanol fuel.

### Gross National Product

Two of the methanol implementation program initiatives could affect the domestic gross national product (GNP). The highway fuel tax would raise the price of all gasoline and some diesel fuel, thus increasing total taxes collected by the federal government and reducing GNP. The methanol price guarantee program would increase GNP through coal-to-methanol plant investment as well as the resultant methanol fuel production.

A highway fuel tax can affect GNP in several ways (some positive, some negative), but the primary effect of such a tax is increased highway fuel costs and correspondingly less funds available for other expenditures, which lowers GNP. Table 10 gives the projected impacts of the $0.40 per gallon tax phased in from 1988 through 1992, assuming that overall new car fuel cost per mile is maintained through the purchase of vehicles with higher fuel economy. It can be seen that the reduced growth would be small in the early years of the tax, but would rise to $9.5 billion by 1992. The impact of a highway fuel price rise on GNP is very dependent upon the nature of the increase, e.g., a $0.40 per gallon increase which was implemented in such a way as to prohibit consumers from adjusting to the higher fuel prices (such as a world oil price shock or an instantaneous tax) could depress GNP by as much as $19 billion in 1992, twice as high as the $9.5 billion impact of a phased-in tax.[4] Such an abrupt price rise would also likely increase the sales of imported automobiles, thus decreasing domestic car sales and GNP even more.

The actual impact of the phased-in $0.40 per gallon highway fuel tax on GNP would likely be less than the values given in table 10. There are two factors to be considered in this regard, both related to how the highway fuel tax revenues collected by the federal government would be used. As discussed earlier, the highway fuel tax would generate net revenues as high as $24.8 billion in 1992. Many economists believe there is a link between high budget deficits and high interest rates and that if all of the tax revenues were used to lower the federal budget deficit, interest rates would fall resulting in increased domestic investment and higher GNP. Alternatively, the government could directly stimulate GNP by simply spending a portion of the tax revenues on

TABLE 10. Projected Effects of the Highway Fuel Tax and Methanol Price Guarantee Initiatives on the Gross National Product ($ Billion)

| | 1988 | 1989 | 1990 | 1991 | 1992 |
|---|---|---|---|---|---|
| Highway fuel tax | −0.4 | −2.6 | −4.7 | −7.2 | −9.5 |
| Methanol price guarantee | 10.1 | 11.9 | 12.3 | 12.3 | 13.1 |
| Total | 9.7 | 9.3 | 7.6 | 5.1 | 3.6 |

goods and services. Tables 9 and 10 show that even assuming a very conservative federal government expenditure multiplier of one (i.e., the expenditure of one dollar by the government increases GNP by one dollar), the federal government could completely offset the GNP impact of the highway fuel tax and still have a significant surplus with which to reduce the federal budget deficit. Stimulating growth through lower budget deficits or by spending part of the tax revenues were not considered in the calculation of the highway fuel tax impacts on GNP in table 10.

More important, however, is the effect of the development of a domestic coal-to-methanol production industry on GNP. Assuming that monies which would otherwise have been used to purchase imported oil were instead used for methanol plant construction and operation (thus increasing the domestic capital pool), GNP would be increased. The projected effects of a 1.5 MBPD domestic methanol production industry with plants coming on line from 1990 through 1994 are shown in table 10.[5] The increased GNP prior to 1990 is due entirely to construction expenditures, it is evenly divided between construction expenditures and methanol fuel value in 1991, and it is dominated by the value of the methanol fuel beginning in 1992. As table 10 shows, the positive effect of the 1.5 MBPD methanol fuel industry on GNP more than offsets the negative impact of the highway fuel tax in each year from 1988 to 1992. The situation would become even more favorable after 1992 as the methanol fuel value would continue to increase (the plants are not all operational until 1994) while the ultimate highway fuel tax level is reached in 1992.

In conclusion, the proposed methanol implementation program would have an overall positive effect on the gross national product as the increased domestic growth associated with the development of a methanol fuel industry would more than offset the negative growth impacts of the highway fuel tax. In addition, the use of the highway fuel tax revenues to lower the federal budget deficit would relieve pressure on interest rates and also stimulate economic growth, though this latter effect cannot be quantified at this time. The positive effects on GNP could be even more favorable if additional methanol plants were constructed in the mid-1990s.

### Consumer Prices and Inflation

The imposition of a highway fuel tax would raise the prices of gasoline and diesel fuel to the automotive consumer. Past rises in highway fuel prices due to the oil price shocks of the 1970s preceded periods of high inflation. Would the proposed highway fuel tax have similar impacts?

The highway fuel tax proposal has been designed to minimize inflationary impacts. By their very nature oil price shocks result in fuel price rises which are both abrupt and unexpected, and which do not facilitate any type of planning whatsoever. On the other hand, the proposed highway fuel tax would be gradually phased in over a five-year period and would be known for two years before the beginning of the phasing-in. Thus, fuel users would have seven years to plan for the maximum level of the tax, which should permit them to avoid many of the possible deleterious impacts. Also, since the highway fuel tax would only affect the prices of gasoline and diesel fuel (not other petroleum products nor even diesel fuel for large commercial trucks), fewer economic sectors of society would be affected by this type of tax compared to general oil price rises or to a total oil tax or imported oil tax.

If it could be shown that (1) consumers will take the future tax levels into consideration when making new vehicle purchases during the seven-year transition to the full tax, with the goal of maintaining equivalent operating costs per mile, and (2) the automotive manufacturers are capable of providing the requisite higher-efficiency vehicles, then a strong case can be made that a highway fuel tax would not necessarily raise operating costs for those consumers who purchase vehicles during the transition period. This would result in higher operating costs and corresponding inflationary pressures only for those consumers who do not change vehicles during this period.

We believe there is considerable evidence that both of the above contentions are valid. Consumer buying preferences over the last ten years have shown that vehicle owners do take fuel economy and operating costs seriously, especially in periods of rising fuel prices. Figure 4 plots the average real fuel cost per mile of each year's new passenger car sales from 1973 through 1983. It can be seen that real fuel costs rose during both the 1973–74 and 1979–80 oil price shocks but fell as soon as consumers were able to adjust by purchasing vehicles with higher fuel economy. The emphasis that consumers have placed on fuel economy is indicated by the fact that overall fuel operating costs have dropped since 1973 even though oil prices have risen. We believe that knowing that the tax would be imposed in future years would encourage most consumers to plan for it by purchasing a more fuel-efficient vehicle. One problem with adjusting to an abrupt highway fuel price change is that it takes several years to turn over the U.S. passenger vehicle fleet. Our proposed tax would allow approximately seven years between enactment of the tax and its full imposition. Most vehicles exchange hands within seven years so such planning could be very widespread.

There is also a large body of evidence that the automotive industry would be able to provide much more fuel-efficient vehicles to the marketplace if there were sufficient demand. Various studies are available which suggest that fleet fuel economies of 40 to 60 miles per gallon (mpg) are feasible with currently available technologies.[6] For example, a price rise of $0.40 per gallon would increase annual gasoline costs by $190 for an owner of a 25-mpg vehicle who drives 12,000 miles per year. This potential fuel cost increase would be completely offset if the owner purchased a 33-mpg vehicle for his or

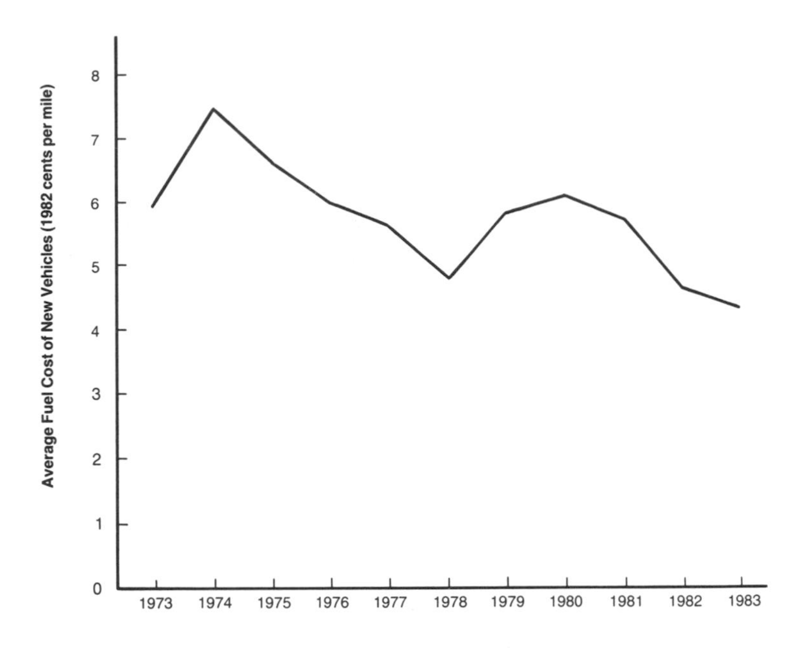


**Fig. 4.** Average real fuel cost per mile of new passenger car sales during each year from 1973 through 1983

her next vehicle purchase. In the long run, the consumer would also have the option of replacing a gasoline vehicle with a methanol vehicle, which would also be likely to reduce some or all of the increased fuel costs due to the tax. Thus, we believe that both the opportunity and desire exists to maintain equivalent fuel costs on the part of individuals purchasing vehicles during the phasing-in of the highway fuel tax.

Nevertheless, not all vehicles turn over within a seven-year time period, and consumers who purchase new vehicles in 1990 or 1991 will still face higher fuel prices as the tax is phased in during 1988 and 1989. Thus, there will be some inflationary pressures due to higher highway fuel prices. Table 11 gives a projection of the impacts of the $0.40 per gallon highway fuel tax on consumer price levels assuming that the tax is phased in from 1988 through 1992 and that overall new car fuel cost per mile is maintained by vehicle purchasers during that period.[7] Two indicators of consumer prices are given. The first set of data indicates that the cumulative impact of the highway fuel tax on overall consumer prices would rise from 0.4 percent in 1988 to 1.5 percent in 1992. The consumer price index (CPI) would be 0.3 to 0.4 points higher each year with the highway fuel tax.

TABLE 11. Projected Impacts of Highway Fuel Tax on Consumer Prices

| | 1988 | 1989 | 1990 | 1991 | 1992 |
|---|---|---|---|---|---|
| Cumulative consumer price levels (1987 = 100) | | | | | |
| Baseline | 104.7 | 109.4 | 114.1 | 119.0 | 124.1 |
| With tax | 105.1 | 110.1 | 115.2 | 120.5 | 125.9 |
| Percentage change | 0.4% | 0.6% | 1.0% | 1.3% | 1.5% |
| Annual consumer price index | | | | | |
| Baseline | 4.7 | 4.5 | 4.3 | 4.3 | 4.3 |
| With tax | 5.1 | 4.9 | 4.7 | 4.6 | 4.6 |
| Points change | 0.4 | 0.4 | 0.4 | 0.3 | 0.3 |

Overall, the methanol implementation program can be seen as insulating the United States from damaging inflationary spirals due to oil price rises. There is always the possibility of a world oil supply disruption causing a third oil price shock, and even in the absence of supply cutoffs world oil prices are expected to rise significantly in the 1990s. By achieving petroleum conservation through both consumer purchases of vehicles with higher fuel economy and the development of methanol as a liquid fuel substitute for petroleum, the United States would largely insulate itself from the economic havoc which would otherwise result from much higher world oil prices. In this context, a small inflationary impact associated with the highway fuel tax can be seen as prudent insurance against inevitable oil price rises and much greater inflationary pressures in the future. On the other hand, because of the capital-intensive nature of methanol plants, and our large domestic coal reserves, methanol would not be expected to be subject to large inflationary pressures once the production plants are in place.

## The Coal Industry

A national coal policy will facilitate an optimum solution to both our acid deposition and imported petroleum problems. This solution involves utilizing low-sulfur coal in electric power plants and converting high-sulfur coal to methanol for use as a transportation fuel. Such a strategy would have significant impacts on both the low-sulfur and high-sulfur coal markets that can only be summarized here.

Allowing the use of low-sulfur coal to be a compliance option for an acid rain control program would result in a large increase in demand for low-sulfur coal. It is impossible to predict the absolute value of this increased demand, as it is dependent upon the details of the specific acid rain program approved, the decisions of scores of individual utilities (utilities would likely install scrubbers on some larger, newer power plants), etc. Projections are that demand for low-sulfur coal would increase by between 90 and 170 million tons per year.[8] This would provide a significant economic stimulus to those areas of the country which contain low-sulfur coal reserves. Our largest low-sulfur coal reserves are in the West, concentrated as subbituminous coal in Wyoming and Montana, but there are also large low-sulfur bituminous reserves in the southern Appalachian coal fields of eastern Kentucky, West Virginia, and Virginia. Thus, these areas would benefit greatly from a program which would increase low-sulfur coal demand.

The high-sulfur coal industry would lose much of its electrical power plant market if an acid rain control program which allows fuel switching is approved by Congress. But the key to our strategy is to maintain demand for high-sulfur coal by utilizing it as a feedstock for methanol production. The methanol implementation program would directly stimulate the production of 1.5 MBPD of methanol from high-sulfur coal, which would consume approximately 130 million tons of high-sulfur bituminous coal annually. This demand would offset the market loss in the utility sector. The employment stimulus of the program in the high-sulfur regions of the Midwest and northern Appalachian regions has already been discussed. It is also important to note the potential growth in high-sulfur coal demand should the initial methanol program stimulate a full-scale movement toward methanol as the primary national transportation fuel. If methanol were to ultimately displace petroleum entirely as a transportation fuel, approximately one billion tons of bituminous coal per year would be needed as feedstock (and a higher total tonnage if subbituminous coal and/or lignite were also used as methanol feedstocks). After construction of the high-sulfur coal-to-methanol plants covered by the price guarantee, it would be expected that many of the second-generation plants would be built in the West to take advantage of lower feedstock costs and to have easy access to West Coast fuel markets. However, we believe plants would continue to be constructed in the East and Midwest as well, in order to be near large population centers in those areas and because water restrictions in the West may limit methanol plant construction there. Thus, the potential exists for large increases in both low-sulfur and high-sulfur coal demands if the methanol implementation program is successful.

### The Automobile Industry

The methanol implementation program could be very beneficial to the American automobile industry. One of the primary reasons for past U.S. dominance of the world automotive market was the recognition of American technological superiority. This position of leadership has deteriorated, however, as first European and then Japanese manufacturers have made major breakthroughs in engine and vehicle design. Alternative fuels provide an opportunity for U.S. manufacturers to regain their leadership position. Methanol is now widely recognized as the most promising future world fuel because it can be produced from a wide variety of feedstocks found throughout the world. Those manufacturers which develop the first successful methanol vehicle designs will accrue a competitive advantage not only in their home markets but potentially in world markets as well. Our proposed program would provide an important stimulus to domestic manufacturers who would then be in a strong position to achieve technological leadership in the emerging world methanol vehicle market.

Although at first thought an unlikely candidate, the highway fuel tax could also have a beneficial impact on the domestic industry. Uncertainty over future automotive fuel prices, due to our dependence on imported petroleum, has been a serious problem for domestic manufacturers. They were caught "shorthanded" during both of the oil price crises of the 1970s, and lost a significant share of the market each time to importers which were manufacturing the more fuel-efficient vehicles that the public suddenly desired. The American automobile industry has realized the importance of fuel-efficient technologies, and has spent considerable resources on design improvements. The question now is what level of fuel economy the market will demand in the future. As long as we rely on imported oil, it is impossible to know with any certainty what fuel economy consumers will demand in the future. As long as future fuel prices are uncertain, any sudden increases in fuel prices will favor importers. Enactment of a known, phased-in highway fuel tax would establish some certainty in future fuel prices, would give assurance to the manufacturers that higher fuel efficiency will be demanded in the future, and would reduce the damage to the industry of another oil price shock.

The methanol vehicle purchase tax credit could also aid the domestic auto industry. Since one fundamental premise of our overall strategy is to stimulate domestic economic growth, perhaps the tax credit should only apply to the purchase of *domestic* methanol vehicles. Doing so would result in greater investment and employment in the U.S. automotive industry and related sectors. It would likely further improve our international trade situation. This option should be seriously considered.

### Air Quality

The most widespread environmental impacts of the methanol implementation program will be related to air quality. Of course, the direct effects will be beneficial—very large reductions in sulfur dioxide emissions from existing utility power plants and lower motor vehicle emissions due to the use of methanol as a substitute for petroleum fuels. In addition to reducing atmospheric acidity levels, which is one of the two primary goals of the strategy, lower sulfur dioxide emissions will also improve compliance with the national air quality standard for sulfur dioxide, and will improve visibility and public health in those areas with high sulfate levels. Compared to gasoline vehicles with advanced emission controls, methanol vehicles will emit lower levels of reactive hydrocarbons and possibly other pollutants as well. Formaldehyde emissions would likely be somewhat higher from methanol vehicles, but catalytic converters are expected to reduce such emissions to low levels. As a substitute for diesel fuel, methanol would provide large emission reductions for a wide range of pollutants, most notably particulate matter, nitrogen oxides, and reactive hydrocarbons. These lower vehicle emissions would be particularly helpful in urban areas which continue to have serious air quality difficulties.

Large coal-to-methanol plants will certainly have the potential for significant environmental impacts. Potential impacts can be greatly minimized at the design stage, however, if stringent control systems are built into the plants. Stack emissions of sulfur dioxide, nitrogen oxides, particulate matter, metals, and various organic compounds would be expected from methanol production plants. One advantage of indirect liquefaction technologies (which go through an interim gasification step) such as methanol production is that most of the organic sulfur and nitrogen is broken down to simpler compounds such as hydrogen sulfide and ammonia. These compounds can be relatively easily separated from the carbon monoxide and hydrogen gases which make up the major parts of the coal gas mixture. In addition, sulfur can deactivate the methanol synthesis catalyst, so there is an economic motivation for its removal. Accordingly, emissions of sulfur and nitrogen oxides would not be expected to be significant. Baghouse filters and electrostatic precipitators have proven to be very effective in controlling particulate emissions and would likely be required for large methanol plants.

Of greater concern would be emissions of various toxic trace metals, such as lead, arsenic, cadmium, mercury, and nickel, and various reactive organic compounds which can be both toxic and very active in the formation of photochemical oxidant. Fortunately, most of the trace metals would be in the form of particles which would be collected either in the coal ash or by the particulate emission control system. Emissions of organic compounds would be much more scattered and problematic and their control should be a prime area for further research. Of course, the fact that coal-to-methanol plants would be located in less densely populated areas is an advantage in terms of overall public exposure.

One final air pollution issue of concern involves carbon dioxide emissions. Some experts believe that increased carbon dioxide emissions will raise worldwide temperatures, and there is concern about major global climatic changes associated with high carbon dioxide emissions. The production and combustion of methanol from coal would raise carbon dioxide emissions relative to gasoline from petroleum (though emissions would be lower than those from vehicles utilizing gasoline from coal). Using current estimates of carbon dioxide emissions from various methanol production processes,[9] and the assumption that methanol vehicles will be 25 percent more energy-efficient than gasoline vehicles, a coal-based methanol system would likely increase carbon dioxide emissions associated with the use of highway vehicles by 55 to 70 percent compared to current vehicles operated on gasoline produced from petroleum.

Fortunately, the potential exists for large improvements in motor vehicle fuel economy as discussed previously in this chapter. Increasing the national fleet fuel economy to 40 or 60 miles per gallon, for example, would decrease fleetwide carbon dioxide emissions even as coal-based methanol-fueled vehicles were being phased into the fleet in the early 1990s. In the long term, if necessary, there exists the possibility of shifting to a biomass-based methanol fuel production system with no net carbon dioxide buildup.

# Part 2 Blueprint for a Transition to a Methanol Future

# CHAPTER 5 Petroleum Tax

In chapter 1 we discussed the U.S. oil import problem in considerable detail. Although net oil imports have fallen from a peak of 8.6 MBPD in 1977 to approximately 4.5 MBPD during the 1982 to 1984 time period, and our oil import bill has decreased from nearly $80 billion in 1980 to $50 billion in 1984 (the drop has been even greater in real dollars), oil imports are still an economic burden on the domestic economy. In particular, they continue to be a primary contribution to our international trade difficulties. We have already seen that our merchandise trade and current account deficits have risen to over $100 billion in 1984, levels that would have been incomprehensible just a few years ago. Many economists believe that the trade deficits portend serious problems for the U.S. economy in the future if they are not reduced.

Yet the outlook for future oil import levels is not bright. Nearly all experts expect domestic petroleum consumption to increase as economic growth continues, and domestic petroleum production to drop as exploration efforts diminish and reserves are depleted. As our levels of oil imports grow, it is expected that the United States will once again become more dependent on OPEC sources, and world oil prices are expected to rise in real terms in the 1990s. Based on current projections, we calculated in chapter 1 that U.S. oil import payments will be approximately $80 billion in 1990 and over $100 billion in 1995 (in 1982 dollars). These levels would seriously worsen our already perilous trade situation. These projections were based on the assumption that no supply constraints would occur; if such a disruption did happen, oil prices would undoubtedly rise further.

Of course, the substitution of methanol for petroleum would greatly ameliorate our oil import and international trade problems. While there is little doubt that long-term world oil prices will rise to such a degree that methanol will ultimately be a more economical vehicle fuel than petroleum, the increasing oil import payments in the interim will continue to undermine our domestic economy. Due to the fluctuating nature (and oligopolistic control) of world oil prices and the time needed for coal-to-methanol plant construction, it may be impossible to begin the methanol transition without accepting some degree of uncertainty. Yet, the country must make such a decision if it is to avoid the continued drain on our economy and the economic havoc associated with major unforeseen oil price shocks.

What is needed is a mechanism that will send a clear signal to the market that petroleum is a fuel whose cost will rise consistently in the future. This is done most directly by the imposition of a petroleum tax, which could take one of many forms. Two specific types of petroleum taxes are analyzed in this chapter: an oil import tax and a highway fuel tax. The adoption of a meaningful oil import or highway fuel tax would convince energy producers and consumers alike that investments in methanol production plants and vehicles are justified on a long-term economic basis. In addition to removing the primary barrier to methanol implementation, uncertainty of future petroleum prices, it will also be shown that a petroleum tax is justified by incorporating a greater portion of the true total cost to the country of relying on imported oil.

One concept that economists have developed to assess the overall impact of large oil import bills is "social cost" or "social premium." The social premium includes those costs to U.S. society of relying on large volumes of imported oil which are not reflected in the market price of imported oil. The total or real cost to U.S. society, then, is the sum of the market and social costs. Although the concept of social cost could be applied to any imported good or service, it is especially appropriate in terms of oil imports because of the critical role of oil in the United States and other industrialized economies, and because the United States itself is such a major oil importer and has assumed responsibility for protecting world oil trade.

Analysts have hypothesized that there are at least four components of the social premium of imported oil. One is the impact of the level of oil imports on the price of world oil. Although the supply/demand relationships of the international oil market are skewed by the OPEC cartel, it is generally accepted that higher U.S. oil imports drive world oil prices upward and lower oil imports drive them downward. More importantly, any increase in price affects all imports, and not just the incremental imports. For example, assume that the United States increases its net oil imports by 2 MBPD, from 4 MBPD to 6 MBPD, and that this increase raises the world oil price by \$3 per barrel. This higher price would be paid for all oil imports, and would increase the U.S. oil import bill by (\$3/barrel) × (6 million barrels/day) × (365 days/year) = \$6.6 billion/year. This \$6.6 billion increase is due to the 2 MBPD increase, and thus the premium for those increased imports is (\$6.6 billion) divided by (2 million barrels/day × 365 days/year) = \$9/barrel. In this example, the increased oil imports are actually costing the United States \$9 per barrel more than their market price, because of the higher price for all oil imports. It can also be argued that the higher world oil price would allow domestic producers to raise their crude oil prices as well. If so, this premium on the increased imports would be correspondingly greater.

A second social cost of imported oil is the impact of the large outlays on our domestic economy—inflation, lower economic growth, unemployment, etc. The evidence strongly indicates that oil imports, especially in times of rising prices and uncertain supply, can have very deleterious macroeconomic effects. Real GNP growth was very low or negative in 1974–75 and 1979–82, both periods directly following oil price shocks. As one analysis indicated, if real GNP growth was retarded by just 1 percent in any of those years due to our reliance on imported oil, it would have involved a loss of $15 billion to $30 billion that year, which would be equivalent to $7 to $14 per barrel of imported oil.[1] Lowering our oil imports would have lessened the macroeconomic impacts and would have lowered the social costs associated with those imports.

A third social cost is the inevitable price rise that would accompany any sudden supply disruption. Again, the magnitude of this social premium is proportional to the level of imports required, and any increase in price brought about by the disruption would affect the price of all oil imports. This component of the imported oil premium is very difficult to estimate.

Finally, there are real costs to the United States associated with trying to militarily ensure a steady supply of international oil, primarily in the Middle East. These costs include direct expenditures for weapons systems and personnel who would be specifically trained for combat in the Middle East (such as the Rapid Deployment Force) and for foreign military assistance to allies in the region. Also important, however, are the costs to our society of the constraints on our freedom of action in foreign policy due to our dependence on some very fragile Middle Eastern governments. Our experiences in Iran and Lebanon should serve to remind us of our vulnerability in dealing with different cultures in that part of the world.

While there is general agreement that all of these factors should be included in the social premium for imported oil, there is no good consensus on the actual values of the various components. One source cited three estimates of the total social premium, ranging from $10 to $72 per barrel.[2] A second source listed eight different estimates derived from various econometric models, with values ranging from $4 to $20 per barrel. These estimates excluded costs associated with the macroeconomic and military expenditure components.[3] The American Petroleum Institute (API) cited six estimates of the total social premium for imported oil, ranging from $2 to $46 per barrel, and concluded that the most likely premium was between $5 and $30 per barrel. The API analysis also excluded the macroeconomic and military expenditure components of the premium.[4] Finally, the Office of Technology Assessment did not estimate a particular value in view of the uncertainties involved, but did claim that "the possible future import premium could range up to $50 per barrel."[5]

## Oil Import Tax and Its Economic Effects

One option to stimulate the production of methanol from coal and to reduce the large direct and indirect costs of imported oil to the United States is the imposition of an oil import tax. A tax on imported oil would alter the relative pricing relationship between imported oil and domestic energy alternatives. If world oil prices were maintained, the oil import tax would stimulate domestic energy development and reduce demand for imported oil, resulting in a smaller outflow of U.S. dollars, increased U.S. employment in energy production, and other associated benefits.

Concern has been expressed that other countries might be able to undercut a domestic coal-to-methanol industry by using natural gas that is currently flared as a methanol feedstock. In order to maintain the domestic economic benefits of the coal-to-methanol transition, the oil import tax should apply, on an energy equivalent basis, to all fossil fuel–derived liquid fuels imported into the United States. Thus, whenever the oil import tax is discussed, its meaning should be taken to include a tax on all crude oil and fossil fuel–derived liquid fuels imported into the United States.

It would be possible to eliminate oil imports entirely by a very large oil import tax, but that would have serious impacts on standards of living, inflation, etc. The following analysis involves three rather modest tax options. It would still be theoretically possible that the oil-exporting countries could undercut domestic energy producers by flooding the U.S. market with abundant, cheap oil. But such major price cutting would be very beneficial to the United States and other industrialized countries, since it would result in significant overall monetary savings. Accordingly, the philosophy behind an oil import tax would not be to eliminate oil imports per se (though that is a possible long-term result), but rather to cap or establish a ceiling for the price of oil imported into the United States. The three options to be examined for an oil import tax are to cap imported oil prices at $40, $30, and $20 per barrel. With world oil prices as of late 1984 at approximately $28 per barrel, these three options cover a broad yet realistic set of alternative pricing scenarios.

Any specific oil import tax proposal must address four basic issues: (1) the level of the tax, (2) the timing of the tax, (3) the duration of the enabling legislation to mandate the tax, and (4) the method to be used in calculating future tax increases with respect to inflation.

The most important factor in determining the level of the tax is that it must be large enough to ensure that methanol production from U.S. coal is stimulated and that imported oil prices would be stabilized. In a previous analysis, EPA has shown that methanol could be sold at retail stations for between $7.60 and $14.30 (in 1981 dollars) per million Btu.[6] This range excludes taxes as will all the methanol and gasoline fuel costs in this analysis.

This would be equivalent to $7.90 to $15.00 per million Btu in 1982 dollars. The Office of Technology Assessment (OTA) has projected that methanol could be sold for between $8.80 and $13.00 per million Btu in 1980 dollars.[7] This would translate to $10.20 to $15.00 per million Btu in 1982 dollars. The EPA and OTA projections are equal at the high end of the range but slightly different at the lower end. Based on these studies, we will use a range of $10.00 to $15.00 per million Btu (1982 dollars) for the future cost of fuel methanol, with an average value of $12.50 per million Btu. The average value of $12.50 per million Btu is equivalent to $0.71 per gallon of methanol.

An oil price of $40 per barrel would yield gasoline at a retail price of $1.34 per gallon or $11.60 per million Btu. Our range of methanol costs of $10.00 to $15.00 per million Btu brackets the cost of gasoline from $40 per barrel crude oil. Using the average of $12.50 per million Btu, methanol would be $0.90 per million Btu more expensive. This is equal to $0.10 per gallon of gasoline or $3 per barrel of crude. A tax of $3 per barrel on imported oil would therefore cap imported oil prices at $40 per barrel, as methanol production, based on our projections, would be more economical at imported oil prices in excess of $40 per barrel. Table 12 shows similar calculations for the $30 per barrel and $20 per barrel scenarios. An oil import tax would have to be at least $13 and $23 per barrel, respectively, to stimulate methanol production and cap imported oil prices at these levels.

The most important aspect of the timing of the tax is that, to as great an extent as possible, it be gradually phased in over a reasonable time frame and not simply implemented in one fell swoop. It is generally accepted that the deleterious macroeconomic impacts of the oil price shocks of 1974 and 1979–80 were due not only to the levels of the price increases but to the swiftness and unpredictability of the increases as well. A gradual and known phasing-in of the tax over a multiyear time period should facilitate considerable planning by petroleum users (and the producers of petroleum-consuming products) and mitigate some of the serious economic impacts that could result from unforeseen price rises.

Beginning the phasing-in of the oil import tax in 1988 would give users ample opportunity to prepare for the tax. The ultimate level of the tax (in real

TABLE 12. Determination of Level of Tax to Cap Imported Oil Prices at Various Levels

| Oil Price ($ per barrel) | Corresponding Price of Gasoline ($ per million Btu) | Projected Cost of Methanol ($ per million Btu) | Difference between Gasoline and Methanol ($ per million Btu) | Requisite Tax on Imported Crude ($ per barrel) |
|---|---|---|---|---|
| 40 | 11.60 | 12.50 | 0.90 | 3 |
| 30 | 8.70 | 12.50 | 3.80 | 13 |
| 20 | 5.80 | 12.50 | 6.70 | 23 |

dollars) should be reached by 1992 to coincide with the phase-out of the methanol vehicle purchase tax credits. The five-year time period from 1988 through 1992 is sufficient for gradually building the tax levels. Table 13 gives the proposed phase-in schedules for the $40, $30, and $20 per barrel oil price cap options.

Related to the issue of timing is the duration of the tax. Petroleum users and methanol producers alike need to know that the oil import tax is here to stay and won't be easily repealed. Thus, the enabling legislation should be written to maintain the tax until well past the turn of the century.

Finally, the tax should be calculated in inflation-adjusted dollars so that the impact of the tax is not diminished over time. This can be easily done by establishing a baseline date (1982 is used here) and calculating each year's tax in the baseline year's dollars. Assuming some level of inflation, the tax would continue to increase after 1992 (though not in real dollars).

The primary justification for an oil import tax is to encourage the development of domestic methanol fuel production that would provide an alternative to imported petroleum. This would lower the direct and indirect economic costs of importing large amounts of crude oil and keep American dollars in this country to promote domestic growth. Several other important effects of an oil import tax must be discussed, however. These include its impact on petroleum conservation and oil import payments, on the federal budget, on the prices and users of various petroleum products, on inflation, and on the concept of equity between various affected groups. It must be noted that the following analyses of the economic impacts of an oil import tax are not as detailed as those presented in chapter 4 for the highway fuel tax (and which will be elaborated upon later in this chapter), as the latter has been incorporated into the methanol implementation program.

### Petroleum Conservation and Oil Import Payments

Oil import fees are justified by some solely on the benefits of the petroleum conservation they would bring about. Conservation in this country has been

TABLE 13. Projected Oil Conservation Impacts and Levels of Oil Imports for $40, $30, and $20 per Barrel Oil Price Cap Options

| | Oil Import Tax (real $ per barrel) | | | Induced Oil Conservation (MBPD) | | | Projected Total Oil Imports (MBPD) | | |
|---|---|---|---|---|---|---|---|---|---|
| Year | $40 | $30 | $20 | $40 | $30 | $20 | $40 | $30 | $20 |
| 1988 | 1 | 2 | 4 | 0.2 | 0.3 | 0.6 | 5.8 | 5.7 | 5.4 |
| 1989 | 1 | 4 | 8 | 0.2 | 0.6 | 1.2 | 5.8 | 5.4 | 4.8 |
| 1990 | 2 | 7 | 13 | 0.3 | 1.0 | 1.9 | 5.7 | 5.0 | 4.1 |
| 1991 | 2 | 10 | 18 | 0.3 | 1.5 | 2.7 | 5.7 | 4.5 | 3.3 |
| 1992 | 3 | 13 | 23 | 0.5 | 1.9 | 3.4 | 5.5 | 4.1 | 2.6 |

very irregular due to the nature of oil prices during the recent past. Many conservation efforts were begun after the oil price rise in 1974, then conservation tailed off in the late 1970s as the real price of fuel leveled off and dropped. It became a national priority again after the 1979–80 price shock, only to be nearly forgotten again in today's petroleum "glut." The advantage of a long-lasting and gradually increasing oil import tax is that it would utilize the marketplace to encourage decision makers to make petroleum conservation a primary concern in their investment policies. In the near term conservation is the very best solution to dependence on imported oil.

Establishment of an oil import tax would encourage petroleum conservation in two ways. First, it would raise the price of imported petroleum (and likely domestic crude oil as well) thus inducing price-based conservation. Second, it would make methanol more competitive with petroleum as a transportation fuel. To whatever degree that methanol displaces gasoline and/or diesel as a motor vehicle fuel, petroleum conservation occurs. In the near term, this latter form of petroleum conservation, displacement by methanol, is dependent, at least to some extent, on the breadth of the methanol price guarantee and vehicle tax credit programs. In the long term, the methanol alternative would become increasingly attractive and ultimately nearly all oil imports would be expected to be displaced by methanol produced from domestic coal feedstocks. The following analysis of petroleum conservation due to an oil import tax includes only that conservation induced by higher prices; it does not include the petroleum conservation which would result from methanol displacement.

The analysis of the conservation impacts of an oil import tax is based on constant American oil consumption of 15 MBPD, a nontax world oil price of $33 per barrel from 1988 through 1992, a constant nontax total import level of 6 MBPD, uses a price elasticity of −0.3 for petroleum products (i.e., demand drops by 0.3 percent for every 1 percent increase in price),[8] and assumes that domestic oil prices would rise to the level of world oil price plus tax, and that all conservation would be reflected in lower oil imports. Table 13 gives the projected levels of petroleum conservation for the various import tax options. The $40 per barrel oil price cap would induce conservation of 0.2 MBPD in 1988 and 0.5 MBPD in 1992. The $30 per barrel cap option would result in reduced U.S. consumption of 0.3 MBPD in 1988, rising to 1.9 MBPD in 1992. The $20 per barrel oil price cap would encourage conservation of 0.6 MBPD in 1988, rising progressively to 3.4 MBPD by 1992. Reductions like these would continue to weaken U.S. dependence on imported oil at a time when demand for oil is likely to grow as economic recovery continues.

The exact effect of an oil import tax on world oil prices is difficult to predict with absolute certainty, especially during the tax phase-in period and before a widespread commitment to construction of coal-to-methanol plants. The assumption of a constant world oil price of $33 per barrel from 1988 through 1992 is probably quite reasonable for the $40 and $30 per barrel cap options, but too high for the $20 per barrel option. The $20 per barrel cap option reflects a $23 per barrel oil tax and would result in $56 per barrel oil by 1992 if the world oil price remained at $33. It is likely that such a tax would exert considerable downward pressure on the price of world oil (from $33 toward $20 per barrel) even during this early transition period. However, since the exact effect is uncertain, a fixed $33 per barrel world oil price is assumed even for the $20 per barrel cap option to facilitate the analysis. Of course, after 1992, the effect of the oil import tax and the methanol implementation initiatives will be to cap world oil prices at the level defined by the tax since any crude oil price increases above the cap level would result in additional large investments in coal-to-methanol plants and the resultant irreversible displacement of petroleum by methanol.

The petroleum conservation induced by the oil import tax would reduce U.S. oil import payments. Assuming that all of the projected conservation levels in table 13 are reflected in lower imports, and that imported oil will cost approximately $33 per barrel during the 1988 to 1992 time frame, table 14 gives the expected reductions in oil import payments. The $40 per barrel cap would reduce oil import payments by $6 billion per year by 1992 while the $30 per barrel cap would result in a $23 billion per year reduction by 1992. The $20 per barrel cap would provide significant reductions, beginning at $7 billion in 1988 and rising progressively to over $40 billion by 1992. Such reductions would significantly improve our international trade balances.

## Federal Budget

Calculating the total impact of an oil import tax on the federal budget is a rather complicated task. Of course the primary impact is the revenue that would be directly raised by the per barrel surcharge. There are several other components to consider when determining the overall effect on federal revenue, however. Because the value of domestic oil would rise to a higher level, the federal government would collect increased windfall profit taxes on oil that was discovered before oil decontrol but which would be sold after de-

TABLE 14. Projected Reductions in Oil Import Payments for the Various Oil Import Tax Scenarios ($ Billion)

| Year | $40 per Barrel Cap Tax | $30 per Barrel Cap Tax | $20 per Barrel Cap Tax |
|---|---|---|---|
| 1988 | 2.4 | 3.6 | 7.2 |
| 1989 | 2.4 | 7.2 | 14 |
| 1990 | 3.6 | 12 | 23 |
| 1991 | 3.6 | 18 | 33 |
| 1992 | 6.0 | 23 | 41 |

control. On the other hand, federal revenue would be reduced somewhat with lower overall macroeconomic activity and profits as a greater percentage of resources were spent on the purchase of petroleum and related products. This would result in lower corporate taxes paid to the federal government. Finally, fewer American dollars would be sent overseas for imported oil and additional funds would be retained in the United States for investment, resulting in higher employment and taxes. The net impact on federal revenue is the sum of all of these individual impacts (as well as other minor impacts which cannot all be predicted).

The Congressional Budget Office (CBO) addressed these issues in an analysis of oil import fees. Analyzing all of the factors listed in the preceding paragraph, it concluded that the net federal revenue generated by an oil import tax was approximately equal to the revenue generated directly by the per barrel tariff. For example, CBO found that at a level of imports of 2 billion barrels per year, oil import taxes of $2 per barrel and $5 per barrel would produce increased net federal revenue of $4.0 billion and $9.9 billion in the near term, respectively.[9] Based on the CBO analysis, it will be assumed that the net federal revenue for the $40, $30, and $20 per barrel oil price cap options would be equal to the direct revenue collected from the tax surcharge.

Table 13 lists the projected levels of oil imports for the years 1988 through 1992 for the $40, $30, and $20 per barrel oil price cap options, respectively, based on the assumptions discussed in the preceding section on conservation. Table 15 gives the projected net federal revenue for these options based on the assumption that the net revenue is equivalent to direct tax revenue. Projections of total oil imports and net federal revenue between 1990 and 1992 must be considered tentative, and would also be very dependent on the magnitude of the methanol price guarantee program. If, as we recommend, a price guarantee program involving 300,000 BPD of additional capacity each year from 1990 through 1994 were adopted, oil import levels and revenue from the tax would both be lower than shown. Projections after 1992 are even more difficult to make as large-scale methanol production would be expected to come on line and begin to significantly displace oil imports.

TABLE 15. Projected Net Federal Revenue for the Various Oil Import Tax Scenarios ($ Billion)

| Year | $40 per Barrel Cap Tax | $30 per Barrel Cap Tax | $20 per Barrel Cap Tax |
|---|---|---|---|
| 1988 | 2.1 | 4.2 | 7.9 |
| 1989 | 2.1 | 7.9 | 14 |
| 1990 | 4.2 | 13 | 19 |
| 1991 | 4.2 | 16 | 22 |
| 1992 | 6.0 | 19 | 22 |

### Higher Prices of Petroleum Products

The most obvious and detrimental impacts of an oil import tax are the higher prices which consumers and manufacturers would have to pay for petroleum products and for goods which utilize oil in their manufacture or transport. Primary petroleum products include gasoline, diesel fuel, fuel oil, jet fuel, and process fuel oil used by industry. The following sections will discuss the impacts of an oil import tax on the prices and users of these products. These analyses will all assume that domestic oil producers will raise the price of their crude oil to the level established by the world oil price plus tax, which would undoubtedly happen until the time when large-scale methanol production would begin. Also, the following analyses will assume that the price increases for petroleum products due to the oil import tax would be proportional to increases in the past when oil prices rose; i.e., we will assume that the relative pricing structure of various petroleum products remains constant. The final section will include a discussion of the likely overall impact of an oil import tax on the inflation rate.

*Gasoline.* Since 43 percent of all crude oil is ultimately refined and used as motor gasoline, the greatest economic impact of an oil import tax would be on consumers with gasoline-fueled vehicles. Based on historical data, we have calculated that every $1 per barrel of an oil import tax would raise the price of a gallon of gasoline by $0.03. Thus, the $40 per barrel oil price cap option, which results in a tax of $3 per barrel in 1990, would increase gasoline prices by $0.09 per gallon. The $13 per barrel tax necessary to cap U.S. oil prices at $30 per barrel would raise gasoline prices by $0.39 per gallon. The $20 per barrel cap tax of $23 would increase gasoline prices by $0.69 per gallon. A person who drives 12,000 miles per year with an automobile that achieves 25 miles per gallon (mpg) requires 480 gallons of gasoline per year. In 1992, this person would have increased gasoline costs of $43, $190, and $330 under the $40, $30, and $20 per barrel oil price cap scenarios, respectively.

Consumers would have the option of purchasing automobiles with better fuel efficiency in order to reduce fuel expenses. The automobile industry has the capability to produce mid-sized vehicles that can achieve much higher fuel efficiencies than available today, and we would expect many consumers to opt for vehicles with higher fuel economy in the face of an oil import tax that was gradually phased in over a several-year time period. Domestic automakers have had a difficult time reacting to the fluctuations in fuel economy needs by the American public in the last ten years, caused by the cyclic nature of world oil prices. An oil import tax could benefit domestic automakers by providing some certainty in the demand for vehicles with high fuel economy. By the late 1980s consumers would also have the alternative of replacing their petroleum-fueled vehicles with methanol-fueled vehicles. While some vehicle owners would undoubtedly have to pay more for gasoline in the short term, the availability of higher fuel economy vehicles and methanol vehicles would give consumers options for decreasing their vehicle operating costs. For example, the $190 annual increase in gasoline costs which would be incurred by the owner of a 25-mpg vehicle with a $13 per barrel oil import tax could be offset if the owner purchased a 33-mpg vehicle instead. Methanol vehicles are expected to be approximately 25 percent more energy-efficient, which would also nearly offset the $190 annual increase. Thus, in the aggregate, we do not expect the oil import tax to have a significant impact on consumer fuel costs, but some individual vehicle owners would experience increases in annual fuel costs.

*Diesel Fuel.* Approximately 9 percent of U.S. crude oil is refined to diesel fuel for use in the commercial trucking industry as well as for most locomotives and an increasing number of diesel cars and small trucks. The impact on diesel fuel prices would be similar to that on gasoline prices, i.e., $0.03 per gallon for every $1 per barrel of tax. Thus, the expected increases would be $0.09, $0.39, and $0.69 per gallon for the $40, $30, and $20 per barrel oil cap options, respectively. The impacts on, and possible solutions for, diesel passenger vehicle owners would be similar to those discussed for gasoline vehicle owners.

The situation would be different for operators of diesel trucks and locomotives. Neither has the option of simply shifting to a more fuel-efficient alternative, though in the long run methanol may prove to be an economical option. Still, higher diesel fuel prices should not significantly impact the freight industries. First of all, trucks and trains are the only major alternatives for most freight shipments, and both would be impacted by higher diesel prices (air freight costs would also be increased by higher jet fuel prices). Second, the economic impacts would not be large. We estimate that fuel costs are approximately 10 percent of the total expenses of trucking, and a lower percentage of overall rail expenses. The $0.39 per gallon tax associated with the $30 per barrel oil price cap would raise diesel prices by approximately 40 percent, and thus overall trucking expenses by approximately 4 percent. If rail expenses rise by a smaller percentage, a small and expected shift in favor of rail freight would be beneficial from a petroleum conservation standpoint, since trains are more energy-efficient than trucks. Third, these costs would be passed on to the consumers of the goods transported by trucks and trains. Thus, in the long run the primary impact of higher diesel fuel prices would show up in higher consumer prices. This issue is discussed in a following section.

*Fuel Oil.* About 5 percent of all U.S. crude is used by the residential and commercial sector for space and water heating. Approximately 12.2 million homes in the United States rely on fuel oil as their primary heating fuel, and the average fuel oil expenditure in these homes in 1981 was $915.[10] Based on an average retail price of $1.15 per gallon, increases of $0.09, $0.39, and $0.69 per gallon in the price of fuel oil would raise this average expenditure by approximately 8, 34, and 60 percent, respectively. These increases would add $73, $310, and $550 to average annual fuel oil expenditures. These increases, in particular the two larger ones, would be especially burdensome in view of the 80 percent increase in average fuel oil expenditure already endured by homeowners from 1978 through 1981.

Some conservation measures are available which could reduce the impacts of higher fuel oil prices, such as switching to nonpetroleum fuels, adding thermal insulation, and otherwise making the home more energy-efficient. But many of these efforts have already been made. For example, 3.6 million households switched from fuel oil to natural gas or wood heating from 1979 through 1981, and the homes which still use fuel oil decreased their volumetric consumption by approximately 20 percent.[11] Thus, there may not be much opportunity to continue to make large energy savings in these homes.

The problem of higher fuel oil costs is exacerbated by the fact that most fuel oil–heated homes are concentrated in one part of the country—the Northeast. Sixty-five percent of all homes using fuel oil for heat are in the Northeast, and these homes use 72 percent of all residential fuel oil consumed in the United States. Nearly 44 percent of all homes in the Northeast use fuel oil.[12] Thus, an oil import tax which increased fuel oil prices, but did not increase the price of other heating fuels such as natural gas or electricity, would place a disproportionate burden on the Northeast. For example, an average annual increase of 34 percent in fuel oil costs (corresponding to the $30 per barrel oil price cap tax) would raise the total annual cost of heating homes in the Northeast by $2.7 billion. The remainder of the country would absorb increased fuel oil costs of only $1.1 billion. These relative impacts make an oil import tax very unappealing to residents and politicians from the Northeast.

*Jet Fuel.* Jet fuel accounts for approximately 7 percent of all U.S. petroleum consumption. The majority of aviation fuel is used for passenger transport, with the remainder used for cargo and personal aviation. As recently as 1973 fuel costs were only 12 percent of airline operating costs, but higher jet fuel prices raised this to approximately 30 percent in 1982.[13] Kerosene-type jet fuel currently sells for about $0.90 per gallon. Thus, increases of $0.09, $0.39, and $0.69 per gallon, corresponding to the $40, $30, and $20 per barrel oil price caps, would raise jet fuel prices by 10, 43, and 77 percent, respectively. These increases would raise overall airline operating costs by 3, 13, and 23 percent, respectively.

If the commercial airline industry were not able to offset these increases, consumer demand could drop significantly as past experience has shown a fairly high price elasticity for flying. This would be good for petroleum conservation, as air travel is more energy-inefficient than other forms of personal transport, but bad for the airline industry. It is likely that the airline industry would be able to ameliorate higher fuel costs. Because fuel costs have historically been a rather small portion of total operating costs, only in the last few years has the airline industry placed a high priority on reducing fuel consumption. Most of the gains in the late 1970s in revenue passenger miles per gallon were the result of short-term improvements in load factor (percentage of seats occupied) and seating capacity (number of seats on plane). It is expected that significant improvements in fuel consumption can be made by introducing new planes with more efficient engines, retrofitting more efficient engines onto existing planes, and improving operating procedures. For example, the Boeing 757 and 767 models consume 35 to 40 percent less fuel per seat mile than the older 727s. Retrofit engines could reduce the fuel consumption of some models by 20 to 30 percent.[14] The development of a short-haul prop-fan aircraft optimized for a somewhat slower cruise speed than now used could lower fuel consumption by up to 50 percent on some routes.[15] Thus, we believe that the passenger airline industry is well positioned to be able to absorb at least a portion of any increased fuel costs due to an oil import tax.

*Industrial Process Fuel.* Our industrial sector accounts for 26 percent of all U.S. petroleum consumption. It uses a wide variety of petroleum products, including liquefied gases, distillate fuel oil, residual fuel oil, ethane, asphalt, etc., both as primary energy sources and as petrochemical feedstocks. On a whole, petroleum products used in the industrial sector tend to be somewhat less valuable than gasoline, diesel fuel, and jet fuel. Thus, for every $1 per barrel increase in the price of oil, we would expect the average price of petroleum products used by industry to increase approximately $0.02 per gallon. Thus, the $3, $13, and $23 per barrel taxes would increase typical petroleum prices by $0.06, $0.26, and $0.46 per gallon, respectively, for the three different oil import tax options.

To the extent that petroleum is consumed by industry to produce goods sold in the United States that do not face stiff competition from imports, any price increases would be passed on to consumers and the primary impact would be a slight increase in inflation. The more critical concern is with respect to the prices and international competitiveness of American exports. Any public policy which threatened to increase our trade deficit even more would have to be viewed with great scrutiny. Some manufacturing processes consume relatively little petroleum and thus would not be affected by an oil import tax—for example, the computer and office machine industry, which is one of our largest exporters. But other industries such as petrochemicals, our second largest exporter, are heavily dependent on petroleum feedstocks. Also, products sold in the United States which have competition from imports and which utilize significant volumes of petroleum in their manufacture could suffer reduced sales. In the long term, even these manufacturers would be able to ameliorate higher energy costs by improving energy utilization efficiency and switching to other fuels such as coal or natural gas. But in the short run it is clear that sales of certain American products, both in U.S. markets and abroad, would be reduced by the imposition of a large oil import tax and that this could result in even higher U.S. trade deficits.

*Overall Impact on Inflation.* The previous sections have briefly outlined the expected price increases for various petroleum products due to an oil import tax and the corresponding impacts on major users. The primary macroeconomic impact of many of these price increases will be on the inflation rate.

It is very difficult to predict the overall inflationary impact of the various oil import tax options. Some increase in inflation is certain, given the increases in the prices of various fuels, but there are several important points to keep in mind. First, the oil import tax can be viewed as an insurance policy against the abrupt oil price shocks we have had in the past that have led to very strong inflationary pressures. A slight inflation rise due to the imposition of a gradually phased-in oil import tax is much easier for our economy to absorb and adjust to than an abrupt and unexpected oil price hike. Second, knowing with certainty that the tax will be gradually phased in over a specific time frame will allow individuals and businesses to plan their investments accordingly, in ways to minimize their oil consumption and thus the inflationary impacts of the higher petroleum prices. Third, as discussed earlier, most energy analysts expect the oil-producing countries to raise the price of world oil once the temporary glut has receded. Thus, to whatever extent that prices set by oil producers are affected by U.S. demand and U.S. demand is lowered by the oil import tax, smaller increases in the price of world oil will result. On the other hand, some of the inflationary impact due to the tax might simply have occurred anyway due to higher world oil price rises.

The most important fact with respect to the higher prices paid by consumers is that the excess capital is being paid to *domestic* corporations who provide jobs and investment capital within the United States and to the federal government to help reduce federal budget deficits. Finally, and probably most important, the investment in methanol production plants that will be stimulated by the oil import tax is strong insurance against future inflationary pressures caused by fuel price increases. Methanol production from coal is very capital-intensive with only about 30 percent of the cost of methanol fuel due to the cost of coal feedstock. Given our huge coal reserves, real coal price increases would not be likely for many years. Therefore, once methanol plants are in place, the price of methanol in real terms should remain nearly constant.

## Equity

There are three equity issues relevant to an oil import tax: (1) will all classes of citizens be treated roughly the same, (2) will all energy companies be affected similarly, and (3) will all regions of the country be treated equitably.

First, almost any energy tax will affect those with lower incomes to a greater degree than it will those with higher incomes. This is because poor families pay a higher percentage of their income for energy than do wealthy families. It is also true that almost all energy used by poor families is non-discretionary (i.e., home heating, trips to work, etc.) and could not be significantly reduced. These factors are partially offset by the fact that poorer families, on the average, drive far fewer miles per year than wealthier families and have smaller homes. The primary benefit to the less advantaged would be the stimulus to domestic energy development and economic growth, which would result in increased domestic employment. Also, poorer families would benefit with other classes from the stabilized U.S. oil prices. Nevertheless, an oil import tax would likely place more of a strain upon the poor than upon other classes of citizens.

There would no doubt be major equity concerns raised by certain oil companies. Those companies with primarily domestic sources of oil would be in the enviable position of being able to raise their prices to the level set by the world oil price plus tax, even though they would not be paying any import tax on domestic oil, thus reaping windfall profits. The windfall profits tax would capture some of these excess profits, but significant amounts would remain. Companies relying on foreign holdings would have no such opportunity, and their competitive positions would weaken in comparison to those of their domestic counterparts. Still, the large oil companies with foreign holdings have shown the ability to survive major structural changes in their businesses before, and it seems unlikely that the oil import tax would seriously injure them. It would be expected that such companies would increase their domestic oil exploration efforts and consider investing in methanol production plant capacity.

Finally, there is the question of regional equity. An oil import tax would be particularly harsh on the Northeast, an area much more heavily dependent on imported oil due to its reliance on fuel oil for heating and residual fuel oil for electrical generation, and because it is distant from domestic oil fields. It has been assumed throughout this analysis, and there is historical evidence to support this assumption, that domestic producers would raise the price of their oil to the level set by the world oil price plus tax. If so, then all oil purchased in the United States would be sold at the new level and New England would not be disproportionately hurt by purchasing more imported oil than domestic oil. Still, since New England relies more heavily on oil, its total energy bill would increase in the near term relative to other regions of the country. It was shown earlier that the Northeast would absorb approximately 70 percent of the entire national cost of higher fuel oil prices. Also, the Northeast consumes much of the residual fuel oil still used for electrical power generation. There is little doubt that the Northeast would be disproportionately impacted by an oil import tax in the near term. In the long term, the Northeast would benefit along with the rest of the country if the oil import tax quickened the development of a domestic methanol industry which would provide an economic alternative to and a price ceiling on imported oil.

### Highway Fuel Tax and Its Economic Effects

The previous section described and analyzed a proposal for an oil import tax. Given that the primary energy problem that the United States faces is the economic impact (both direct and indirect) of our dependence on imported petroleum, an oil import tax can be seen as the most appropriate and efficient mechanism for solving our problem—it would raise the effective price of the specific commodity that we wish to conserve, thereby encouraging conservation in every consumption sector. An oil import tax would also stimulate domestic oil exploration and production, since domestic oil would be more valuable, thereby lessening the need for imported oil even more. Thus, in the most fundamental sense, an oil import tax seems to be the most desirable type of energy tax.

Unfortunately, however, our analysis of the impacts of an oil import tax revealed several drawbacks associated with such a tax: the cost of goods that utilize significant amounts of petroleum in their manufacture or transport could be increased substantially, affecting the competitiveness of both American exports abroad and homemade products in competition with imports; the higher cost of fuel oil used for space and water heating would have a disproportionate impact on homeowners in the northeastern United States; certain oil companies with domestic reserves would reap very large windfall profits, only part of which would be recovered through the windfall profits tax; and the higher cost of diesel fuel would increase expenses for the nation's intercity trucking industry. It could be argued that the importance of each of these individual problems is low compared to the overriding need to reduce our economic dependence on foreign oil, and, moreover, that policies could be adopted to solve or ameliorate all of these concerns. Nevertheless, each of these issues would be expected to spark powerful political opposition (the petrochemical industry, the entire New England political structure, consumer groups opposed to windfall profits, the trucking industry, etc.) which could doom an oil import tax proposal before it was even seriously debated.

In view of the probable political opposition to oil import taxes, it is prudent to consider other alternatives that could produce many of the same benefits with fewer of the drawbacks. The most promising alternative to an oil import tax would be a highway fuel tax. The highway fuel tax we are describing is similar to a gasoline tax except that in general both gasoline and diesel fuels would be included, although certain commercial uses would be exempted. Similar to the oil import tax option discussed above, the highway fuel tax should apply, on an energy equivalent basis, to all fossil fuel–derived highway fuels imported into the United States, including imported methanol.

Although the impact on most motor vehicle operators would be the same, the highway fuel tax would differ from the current federal motor fuel excise tax in two ways. First, while the basis of the current excise tax is to raise revenue specifically for the construction and maintenance of roadways (and, to a lesser degree, the operation of mass transit systems), the primary reasons for a new highway fuel tax are to encourage the use of methanol as an alternative to petroleum, reduce oil imports, and raise revenue to reduce the federal budget deficit. Second, in addition to those public vehicles currently exempted from the federal motor fuel excise tax, the highway fuel tax would allow certain commercial vehicles to be exempted as well. The underlying philosophy would be to exempt those commercial trucks transporting raw materials and components used in the production of finished goods that are exported or in competition with imports. Taxing the fuel used by these trucks would raise the price of the finished product, placing the U.S. manufacturer in a reduced competitive position relative to imports. While the specific details governing the availability of this exemption could be examined further, we recommend that the exemption be granted only to trucks with a Gross Vehicle Weight Rating (GVWR) in excess of 50,000 pounds, since these are the trucks primarily used for this purpose.

While a highway fuel tax would not encourage conservation of all petroleum products, it would maintain the incentive to conserve in the transportation sector, which consumes the largest fraction of the crude oil barrel. Highway transportation consumption of gasoline and diesel fuel accounts for approximately 50 percent of all U.S. petroleum consumption.[16] Unlike an oil import tax, a highway fuel tax would not be expected to raise the cost of fuel to industry, homeowners, or airlines, and would not raise the value of domestic petroleum holdings. Some truckers would be affected, but a large fraction of the intercity trucking industry would not. Thus, such a tax would avoid many of the political battles that could derail an oil import proposal while encouraging conservation in the most important petroleum-consuming sector—highway transport. This is particularly relevant because it is precisely in the highway transportation sector that methanol will ultimately provide an economical alternative to petroleum. Thus, in terms of methanol displacement of petroleum, a highway fuel tax is well targeted. Finally, like an oil import tax, a highway fuel tax will generate revenue that can be used to lower the federal budget deficit. In fact, the highway fuel tax is a more efficient revenue raiser in that all of the tax will be realized by the federal treasury; some of the consumer cost of an oil import tax would be absorbed by oil companies in the form of windfall profits.

Two ultimate levels for a highway fuel tax will be considered: $0.40 and $0.70 per gallon. These levels are approximately equal to the expected increases in gasoline and diesel fuel prices due to the imposition of an oil import tax designed to cap U.S. oil import prices at $30 and $20 per barrel, respectively. For reference, it should be noted that several Western European countries, including France, Germany, and Italy, have gasoline taxes between $1 and $2 per gallon.[17] Currently, the U.S. federal motor fuel excise tax is $0.09 per gallon, with states adding approximately $0.10 per gallon more. The addition of the new highway fuel tax would not affect current federal and state motor fuel excise taxes.

Like an oil import tax, a highway fuel tax should be gradually phased in over a multiyear period. This would give individuals and businesses considerable opportunity to plan for higher gasoline and diesel fuel prices. If, for example, a highway fuel tax were enacted by Congress in early 1986, and phased in from 1988 through 1992, users would have seven years to plan before being faced with the maximum tax. Since most vehicles are turned over within seven years, most individuals and fleet operators would be able to adjust for higher prices through changes in vehicle purchasing decisions. Table 16 proposes phase-in schedules for the two highway fuel tax options. Also, as discussed earlier, the tax should be calculated in inflation-adjusted dollars so that the tax is not eroded by inflation, and the enabling legislation for the tax should be written to maintain the tax until well into the next century so that everyone recognizes that the tax is permanent and not easily repealed.

TABLE 16. Proposed Schedules for the Highway Fuel Tax Options (Real $ per Gallon)

| Year | $0.40 per Gallon Tax | $0.70 per Gallon Tax |
|---|---|---|
| 1988 | 0.08 | 0.14 |
| 1989 | 0.16 | 0.28 |
| 1990 | 0.24 | 0.42 |
| 1991 | 0.32 | 0.56 |
| 1992 | 0.40 | 0.70 |

A highway fuel tax could have significant impacts on petroleum conservation and oil import payments, the federal budget, the gross national product, consumer prices and inflation, and equity. These impacts will be examined in the following sections. Much of the analysis relies on the results of a study of the overall macroeconomic impacts of highway fuel taxes conducted with Jack Faucett Associates, Inc., an economics consulting firm. This analysis projected the macroeconomic effects of twelve different highway fuel taxes, including two levels ($0.40 and $0.70 per gallon), two types of implementation schedules (a gradual phasing-in and a one-time shock), and three "fuel economy adjustment" scenarios (varying assumptions regarding the extent to which consumers would purchase vehicles with better fuel efficiency in response to the tax). The following sections will present results for the $0.40 and $0.70 per gallon highway fuel taxes phased in over a five-year time period. Ranges will be used to bracket the probable consumer reactions to the tax. At one extreme we will assume that consumers will purchase vehicles with much higher fuel economy so as to maintain a constant overall new car fuel cost per mile during the phasing-in time frame. At the other extreme we will assume that consumers only purchase increased fuel economy to the extent that the increase in new car fuel cost per mile due to the tax is one-half of what it would have been if no changes in purchasing decisions were made. A report detailing the results of all of the tax scenarios has been published.[18]

### Petroleum Conservation and Oil Import Payments

A highway fuel tax would induce conservation of gasoline and diesel fuel only. Nevertheless, these highway fuels account for one-half of all U.S. petroleum consumption. The exemption for large trucks that ship raw materials will exclude some diesel fuel from the tax, but still approximately 45 percent of all petroleum used today would be affected by the tax.

The imposition of a highway fuel tax would encourage gasoline and diesel fuel conservation in two ways: it would raise the prices of these fuels resulting in price-induced conservation, and it would make methanol more competitive with petroleum fuels for use in motor vehicles. Price-induced conservation would be most important in the near term, with methanol displacement becoming more critical in the long term. As with our earlier analysis of an oil import tax, we will be able to quantify only the near-term conservation impacts of the highway fuel tax due to higher gasoline and diesel fuel prices.

Table 17 gives the projected petroleum conservation induced by the $0.40 per gallon and $0.70 per gallon highway fuel taxes, respectively. These values were derived in the study described earlier, and are based on the following assumptions: gasoline and diesel fuel consumption by nonexempted

TABLE 17. Projected Conservation Impacts and Levels of Highway Fuel Consumption for $0.40 and $0.70 per Gallon Highway Fuel Taxes

| | Highway Fuel Tax (real $ per gallon) | | Induced Highway Fuel Conservation (MBPD) | | Projected Highway Fuel Consumption (MBPD) | |
|---|---|---|---|---|---|---|
| Year | $0.40 | $0.70 | $0.40 | $0.70 | $0.40 | $0.70 |
| 1988 | 0.08 | 0.14 | 0.24–0.38 | 0.37–0.59 | 5.62–5.76 | 5.41–5.63 |
| 1989 | 0.16 | 0.28 | 0.38–0.60 | 0.60–0.94 | 5.40–5.62 | 5.06–5.40 |
| 1990 | 0.24 | 0.42 | 0.52–0.79 | 0.84–1.22 | 5.21–5.48 | 4.78–5.16 |
| 1991 | 0.32 | 0.56 | 0.65–0.93 | 1.05–1.44 | 5.07–5.35 | 4.56–4.95 |
| 1992 | 0.40 | 0.70 | 0.79–1.07 | 1.26–1.65 | 4.93–5.21 | 4.35–4.74 |

vehicles would be 6.0 MBPD during the 1988 to 1992 time frame; crude oil prices would be approximately $33 per barrel during these same years, excluding the new tax; the price elasticity of gasoline and diesel fuel is −0.3; refiners would maintain their present petroleum products pricing structures; methanol displacement of petroleum fuels would remain small until after 1992; and consumers would modify purchasing decisions so as to offset increased new car fuel cost per mile by somewhere between 50 and 100 percent.[19] Table 17 shows that the phasing-in of the $0.40 per gallon tax would promote conservation of 0.24 to 0.38 MBPD in 1988, rising progressively to 0.79 to 1.07 MBPD by the time the ultimate tax level were reached in 1992. Of course, conservation would be considerably greater under the $0.70 per gallon tax, peaking at 1.26 to 1.65 MBPD in 1992 as shown in table 17. These large reductions in highway fuel consumption might require refiners to modify their pricing structures in order to continue to be able to market a feasible slate of petroleum products. For example, they might lower gasoline prices (to increase demand) and raise home heating fuel or residual fuel prices. We cannot project the likelihood or consequences of such an action, but overall petroleum conservation would not be significantly changed as most petroleum products have very similar price elasticities.[20]

The large reductions in petroleum consumption brought about by a highway fuel tax could significantly reduce U.S. oil import payments. Table 18 gives the projected improvements in oil import payments for the $0.40 and $0.70 per gallon highway fuel taxes. These projections are taken from the macroeconomic analysis of highway fuel taxes, and are based on assumptions that the $0.40 and $0.70 per gallon highway fuel taxes encourage petroleum conservation at the levels shown in table 17, and that imported oil will cost approximately $33 per barrel in the late 1980s and early 1990s.[21] It can be seen in table 18 that the price-induced conservation promoted by the $0.40 per gallon tax could reduce U.S. oil import payments by $10.3 billion to $13.7 billion per year by 1992. The $0.70 per gallon tax would induce higher levels

TABLE 18. Projected Reductions in Oil Import Payments for the $0.40 and $0.70 per Gallon Highway Fuel Taxes ($ Billion)

| Year | $0.40 per Gallon Tax | $0.70 per Gallon Tax |
|---|---|---|
| 1988 | 3.0–4.8 | 4.8–7.5 |
| 1989 | 4.9–7.6 | 7.8–11.8 |
| 1990 | 6.8–10.0 | 11.0–15.5 |
| 1991 | 8.6–11.8 | 13.7–18.3 |
| 1992 | 10.3–13.7 | 16.4–21.1 |

of conservation, of course, and would be expected to improve oil import payments by between $16.4 billion and $21.1 billion annually by 1992. Implementation of a highway fuel tax is one promising strategy to reduce our trade deficits by $10 billion to $20 billion per year.

This analysis of conservation and oil import payments has been based on the assumption that U.S. oil prices would be approximately $33 per barrel during the 1988 to 1992 time frame. This seems particularly appropriate for the $0.40 per gallon highway fuel tax, which would be expected to cap U.S. oil prices at $30 per barrel by 1992 when the tax is completely phased in and methanol production is under way. Since the $0.70 per gallon tax would ultimately cap U.S. oil prices at $20 per barrel, it is possible that enactment of this tax would exert downward pressure on U.S. oil prices even during the tax phase-in period. If so, U.S. oil prices under the $0.70 per gallon tax might be below $33 per barrel, which would result in slightly different conservation and oil import impacts.

## Federal Budget

As discussed earlier, it is very difficult to accurately project the overall impact of an energy tax on the federal budget. Such an analysis of a highway fuel tax is somewhat simpler than that of an oil import tax, since fewer economic sectors are affected and the windfall profits tax is not a factor. The primary impact of a highway fuel tax is the direct revenue generated by the tax. This is a very straightforward determination given the level of the tax and projected future highway fuel consumption. This increased federal revenue will be partially offset by lower taxes paid to the government due to lower overall macroeconomic activity and profits, as more resources would be used for the purchase of gasoline and diesel fuel.

Projected increases in overall federal revenue generated by the $0.40 and $0.70 per gallon highway fuel taxes are shown in table 19.[22] The $0.40 per gallon tax would increase federal revenue by over $6 billion in 1988, by over $17 billion in 1990, and by approximately $27 billion in 1992 and subsequent

TABLE 19. Projected Net Federal Revenue Increases for the $0.40 and $0.70 per Gallon Highway Fuel Tax Scenarios ($ Billion)

| Year | $0.40 per Gallon Tax | $0.70 per Gallon Tax |
|---|---|---|
| 1988 | 6.3 | 10.6 |
| 1989 | 11.9–12.1 | 19.4–20.1 |
| 1990 | 17.2–17.8 | 27.5–29.1 |
| 1991 | 22.2–23.0 | 34.9–37.2 |
| 1992 | 26.9–28.0 | 41.6–44.5 |

years. The $0.70 per gallon tax would add over $10 billion to the federal treasury in 1988, rising to approximately $28 billion in 1990 and to over $40 billion in 1992 and later years.

These large increases in federal revenue are major benefits of a highway fuel tax. As discussed in chapter 4, current projections are that annual federal budget deficits will remain in the $200 billion range for at least several years. Some experts believe the deficit will be even higher during the late 1980s. The highway fuel tax could offset a significant part of our overall federal budget deficit, and be an important step in showing the sincerity of our political leaders in resolving the deficit problem. Although increased taxes are never popular, we believe the imposition of a highway fuel tax is politically feasible for two reasons. First, it would be seen as an integral part of the overall methanol implementation program to stimulate domestic energy production and employment and to decrease the export of American dollars. Second, the U.S. public has traditionally preferred taxes on consumption to other types of levies (such as income or property taxes). Thus, a highway fuel tax may well be the most politically appealing way to reduce our federal budget deficits.

### Gross National Product

A highway fuel tax could affect the gross national product (GNP) in several ways. The predominant impact is that the increased level of taxes collected by the government will cause overall income and personal consumption expenditures to decline, thus decreasing GNP. The reduction in oil imports discussed above will tend to increase GNP, as would lower overall imports due to slightly lower funds available for general consumption, but these positive impacts would not be sufficient to offset the negative GNP impacts.

Our macroeconomic analysis calculated the GNP impacts of various types of highway fuel taxes. Table 20 presents the results of the analysis for $0.40 and $0.70 per gallon highway fuel taxes phased in during the 1988 to 1992 time frame.[23] The ranges represent different impacts depending upon assumptions regarding fuel economy purchasing decisions. The $0.40 per

TABLE 20. Projected Reductions in the Gross National Product for the $0.40 and $0.70 per Gallon Highway Fuel Taxes ($ Billion)

| Year | $0.40 per Gallon Tax | $0.70 per Gallon Tax |
|---|---|---|
| 1988 | 0.4–2.5 | 1.3–4.6 |
| 1989 | 2.6–5.8 | 4.9–10.3 |
| 1990 | 4.7–8.8 | 8.0–14.9 |
| 1991 | 7.2–11.8 | 11.8–19.6 |
| 1992 | 9.5–14.6 | 14.7–23.5 |

gallon tax would reduce GNP by $0.4 billion to $2.5 billion in 1988, and by $9.5 billion to $14.6 billion in 1992. The GNP impacts of the $0.70 per gallon tax would be larger, of course, amounting to $1.3 billion to $4.6 billion in 1988 and rising to between $14.7 billion and $23.5 billion by 1992. One of the most noteworthy conclusions of the analysis is that the impact of a highway fuel price rise on GNP is very dependent upon the nature of the increase; for example, a price increase implemented in such a way as to prohibit consumers from planning for the higher fuel prices (such as a world oil price shock or an "instantaneous" tax) would depress GNP much more than a similar price rise phased in over a multiyear period. An abrupt fuel price rise would also be likely to increase the sales of imported automobiles, which would decrease domestic car sales and lower GNP even more. Thus, this analysis supports the contention that a gradually phased-in tax is much more favorable than an unforeseen price increase.

It must be emphasized that the projected GNP impacts shown in table 20 represent worst-case conditions, and in fact are very unlikely to be realistic. There are two factors which must be considered in this regard, both related to how the highway fuel tax revenues collected by the federal government would be used. As shown in table 19, the highway fuel tax would generate large new revenues for the federal treasury. Many economists believe there is a link between high budget deficits and high interest rates, and that if all of the tax revenues were used to lower the federal budget deficit, then interest rates would fall resulting in increased domestic investment and a higher GNP. Alternatively, the government could directly stimulate GNP by simply spending a portion of the tax revenue on goods and services. Tables 19 and 20 show that even assuming a very conservative federal government expenditure multiplier of one (i.e., the expenditure of one dollar by the government increases GNP by one dollar), the federal government could completely offset the GNP impact of the highway fuel tax and still have a significant surplus with which to reduce the federal budget deficit. Assuming a federal government expenditure multiplier greater than one would result in an even greater surplus for reducing the budget deficit. Stimulating growth through lower budget deficits or by spending part of the tax revenues were not considered in the calculation of the highway fuel tax impacts on GNP given in table 20.

### Consumer Prices and Inflation

The imposition of a highway fuel tax would raise the prices of gasoline and diesel fuel to the automotive consumer. Past rises in highway fuel prices due to the oil price shocks of the 1970s preceded periods of high inflation. This highway fuel tax proposal has been designed to minimize such impacts. The tax would be gradually phased in over a five-year period and would be known to the general public for up to two years before any tax would be levied. Thus, fuel users would have up to seven years to plan for the maximum level of the tax, which should permit most people to alleviate many of the possible deleterious impacts. Since the highway fuel tax would affect only the prices of gasoline and diesel fuel (and not the prices of other petroleum products nor even the price of diesel fuel for large commercial trucks), many fewer economic sectors of society would be affected by this type of tax compared to general oil price rises, a total oil tax, or an imported oil tax.

It is helpful to consider the impact of the highway fuel tax on two groups of consumers—those who purchase vehicles prior to the tax reaching its ultimate level, and those who do not purchase vehicles in that time frame. Evidence of the desire of the former group to maintain equivalent fuel costs per mile by purchasing more efficient vehicles (see fig. 4 showing consumer reactions to previous oil price rises) and of the ability of domestic manufacturers to provide higher efficiency vehicles was presented in chapter 4. For example, a price rise of $0.40 per gallon would increase annual gasoline costs by $190 for an owner of a 25-mpg vehicle who drives 12,000 miles per year. This potential fuel cost increase would be completely offset if the owner purchased a 33-mpg vehicle for his or her next vehicle purchase. In the long run, the consumer would also have the option of replacing a gasoline vehicle with a methanol vehicle, further reducing some or all of the increased fuel costs due to the tax. Thus, we believe that both the opportunity and desire exists to maintain equivalent fuel costs on the part of most individuals who purchase vehicles during the phasing-in of the highway fuel tax.

Nevertheless, not all vehicles turn over within a seven-year time period and consumers who wait to purchase new vehicles in 1990 or 1991 will still face higher fuel prices as the tax is phased in during 1988 and 1989. Thus, there will be some inflationary pressures due to higher highway fuel prices. Table 21 gives our projection of the impacts of the $0.40 and $0.70 per gallon highway fuel taxes on consumer price levels assuming that the taxes are phased in from 1988 through 1992.[24] The consumer price index (CPI) would be 0.3 to 0.4 points higher each year with the $0.40 per gallon

TABLE 21. Projected Annual Increases in the Consumer Price Index for the $0.40 and $0.70 per Gallon Highway Fuel Taxes

| Year | $0.40 per Gallon Tax | $0.70 per Gallon Tax |
|---|---|---|
| 1988 | 0.4 | 0.7 |
| 1989 | 0.4 | 0.6–0.7 |
| 1990 | 0.3–0.4 | 0.6 |
| 1991 | 0.3 | 0.5–0.6 |
| 1992 | 0.3 | 0.5 |

*Note:* Values are in absolute percentage points. Annual baseline CPI values are expected to be between 4.3 and 4.7 during this time frame.

highway fuel tax and 0.5 to 0.7 points higher each year with the $0.70 per gallon tax. Baseline CPI values are expected to rise between 4.3 and 4.7 points each year in absence of any fuel tax. The analysis also showed that some vehicle owners would probably travel fewer miles in order to reduce total fuel costs.

Two other issues need to be considered in this regard, however. First, the market basket of goods upon which the CPI is based is modified every few years to reflect changing consumer purchasing patterns. Since it is not possible to project how and when the CPI market basket will be changed, the data in table 21 are based on the assumption that consumers will continue to purchase as much fuel in the late 1980s and early 1990s as they do today. If, as expected, consumers purchase vehicles with higher fuel economy in response to the highway fuel tax, then they will require less fuel which will reduce the increase in the CPI due to the imposition of a highway fuel tax. Second, there is always the possibility of a world oil supply disruption causing a third oil price shock, and even in the absence of supply cutoffs world oil prices are expected to rise significantly in the 1990s. By achieving petroleum conservation through both consumer purchases of vehicles with higher fuel economy and the development of methanol as a liquid fuel substitute for petroleum, the United States would largely insulate itself from the economic havoc that would otherwise result from much higher world oil prices. In this context, the small inflationary impact associated with the highway fuel tax can be seen as prudent insurance against inevitable oil price rises and much more severe inflationary hikes in the future.

## Equity

A highway fuel tax raises two issues with respect to equity: (1) inequality of the tax burden on various socioeconomic classes, and (2) relative impacts on different regions of the United States.

It is well known that excise taxes are generally regressive in nature. This is because families with lower incomes nearly always spend greater proportions of their budgets on those items to be taxed (generally, necessities) than do families with larger incomes. A highway fuel tax may be an exception to this rule, however. Vehicle ownership is very dependent on income. As of 1977, the last year for which statistics are available, the average number of passenger vehicles per household was 0.73 for families with incomes less than $5,000 and 1.27 for families with incomes between $5,000 and $10,000. Alternately, families with incomes above $25,000 on average owned 2.43 vehicles per household.[25] Annual vehicle miles driven per vehicle is also income-dependent, though to a somewhat lesser degree. Thus, lower-income families own fewer vehicles and drive those vehicles fewer miles. Families with incomes below $5,000 (including those without cars) averaged just 4,500 miles in 1977, while families with incomes between $5,000 and $10,000 averaged 10,300 miles. For comparison, families with incomes over $25,000 drove nearly 26,000 miles in 1977.[26] Thus, on average, upper-income households drove two and a half times as many miles in 1977 as households in the $5,000 to $10,000 income class and nearly six times as many miles as families making less than $5,000 per year. Total fuel costs would not be in the same exact proportion, as it would be likely that wealthier families own newer and more fuel-efficient vehicles. But clearly, wealthier households do pay far more dollars per year for automotive fuel than do poorer families, and in a proportion fairly similar to their after-tax income advantage over low-income families.

While it appears that a highway fuel tax would not be highly regressive with respect to overall socioeconomic equity, it is undoubtedly true that certain individual low-income households would be adversely affected. This is particularly clear with respect to the very poorest families that own and operate a motor vehicle. One reason why households with less than $5,000 in income drove so few miles was that approximately half of these families did not own any personal vehicles whatsoever. While the under-$5,000 households averaged only 4,500 miles apiece, those families that did own a vehicle drove on average over 7,000 miles. In recognition of the fact that poorer households often own vehicles with very low fuel economy, it is likely that low-income vehicle-owning households would lose a greater proportion of income to a highway fuel tax than any other income class.

The regional impact issue revolves around the fact that certain states and regions are far more dependent on personal transport than other areas of the country. For example, in 1975 state highway gasoline sales per household ranged from 864 gallons in New York to 2,222 gallons in Wyoming.[27] The primary reasons for such differences in consumption are the spatial and urban/rural characters of the various states. A congressional analysis has confirmed the expectation that states such as Wyoming, Nevada, Montana, and other western states would be especially hard hit by a gasoline tax. Small and/or highly urbanized states like New York, Rhode Island, and Massachusetts would not be affected as adversely.[28] Rural dwellers would be particularly affected by a highway fuel tax. If necessary, it would be possible to formulate tax policy so as to give low-income households and rural dwellers some relief from the highway fuel tax.

## Conclusion

This chapter has proposed several petroleum tax options, differing not only by whether the tax would apply to imported petroleum or all highway fuels, but also by varying magnitudes of the tax. Oil import tax levels of $3, $13, and $23 per barrel and highway fuel taxes of $0.40 and $0.70 per gallon (which correspond to the two larger oil import taxes) have been considered. The major economic impacts of each of these options have been discussed.

Strictly on the merits, the oil import tax is very attractive. Given current energy prices, the $13 per barrel oil import tax would be preferred since it would cap U.S. oil import prices at approximately $30 per barrel. The economic impacts of an oil import tax on U.S. manufacturers that depend heavily on exports and on the domestic trucking industry, and equity concerns with respect to residents of the Northeast and the oil industry, however, would make political acceptance of an oil import tax very difficult.

The $0.40 per gallon highway fuel tax is recommended at this time. A highway fuel tax is well targeted since highway fuels comprise a significant portion of all petroleum consumption and methanol would be competitive with highway fuels before it would be competitive with other petroleum products. Only the prices of gasoline and diesel fuel would be affected, and because of the potential for improved fuel economy most vehicle owners would be able to reduce the impacts of higher fuel prices. The $0.40 per gallon level is preferred since it would cap U.S. oil prices at approximately $30 per barrel in the long term (equivalent to approximately $1.00 per gallon gasoline excluding all fuel taxes and approximately $1.20 per gallon including present state and federal taxes), which is slightly above current world oil prices.

# CHAPTER 6 Methanol Price Guarantee

The biggest obstacle blocking movement toward a national methanol network is the reluctance of the private sector to invest the billions of dollars of risk capital in large coal-to-methanol plants necessary to produce the quantities of methanol that would be needed to fuel a significant portion of our transportation system. This private sector reticence is understandable in light of the magnitude of the investments involved, the uncertainties surrounding future world oil prices, and the questions concerning the marketability of massive amounts of a transportation fuel that is significantly different from the hydrocarbon fuels marketed today. In particular, potential methanol producers fear that the oil-producing countries might underprice methanol just as it was being introduced to the market and beginning to produce return on investment. While this would be very beneficial to the United States and other industrialized countries in terms of lower oil import bills, it would be disastrous for methanol producers, who would face major economic losses.

In view of these concerns, it would be surprising indeed for large amounts of risk capital to be invested, *even if* all indicators showed that methanol could be produced in the future at a lower equivalent cost than that projected for crude oil imports. Of course, this problem is magnified by the current temporary glut of world oil and falling petroleum prices. Given that development of a national methanol production and distribution system would provide significant economic benefits in the long run (as protection against future oil price hikes and/or supply disruptions), as well as contributing to national security, energy independence, environmental protection, and domestic employment, and that the private sector appears unwilling to accept the investment risks inherent in the development of first-generation methanol production plants, there exists a classic justification for governmental action.

While a government role in the construction of first-generation methanol production plants appears necessary, it is important to recognize that this role should be limited by several important considerations. These include: (1) minimum federal involvement in the marketplace, i.e., relying on the private sector to make as many of the important decisions as possible; (2) minimum cost to the taxpayer, by allowing as much economic self-sufficiency to the industry as possible; (3) least uncertainty of future costs to the government; and (4) maximum ability of public and private decision makers to identify the full detailed cost performance of the production processes.[1] Thus, the appropriate role of government is *not* to create a long-lasting government-dominated, permanently subsidized methanol production infrastructure, but rather to provide just enough financial incentive, in a straightforward and temporary way, that the private sector will make the necessary investment in the first generation of commercial methanol plants. The market mechanism must be relied upon to as great a degree as possible to ensure maximum economic efficiency of methanol production.

There are a number of financial incentives that can be utilized by government to encourage methanol plant investment. The major types of incentives include the following: tax benefits (such as investment tax credits or accelerated depreciation), direct government construction grants, low-interest loans or government-backed loan guarantees, competitive fixed price and non-negotiated open-end price guarantees, and production subsidies other than price guarantees, such as direct cash outlays or production tax credits. Each of these incentive types has both advantages and disadvantages and the appropriateness of each incentive can vary depending on the design of an individual alternative fuel project.

Many have suggested that the proper course for the government is to offer a wide range of incentives so that individual corporations or consortia may choose the most appropriate incentive or group of incentives for its project. This is, for example, the course chosen by Congress in the Energy Security Act of 1980. That law established the Synthetic Fuels Corporation and authorized a wide range of incentives that the corporation could use in assisting various types of synthetic fuels projects.[2] But allowing a variety of incentives has some significant drawbacks: it fails to resolve the question of which type of incentive maximizes social benefit at minimal governmental risk and cost, it results in both industry and government having to establish bureaucracies of analysts in order to determine which incentive(s) ought to be used in any particular project, and it makes it much more difficult for private decision makers and the general public to see the actual subsidies being given and to thus surmise the overall attractiveness of similar future private ventures. For these reasons, it seems most appropriate to focus on one financial incentive program which would simplify the incentive process and facilitate public and private understanding of the technical feasibility and economic performance of an individual project.

In order to identify the most appropriate financial incentive to encourage private sector involvement in methanol production, it is necessary to determine the barriers which now prohibit such investment. Four important barriers to investment in synthetic fuels plants have been identified. One is simply the *economic feasibility* of a synthetic fuel plant. Without any such plants in existence, it is not known exactly what the final cost of the product would be. Such fuels might be competitive with conventional energy sources, but they might not be. This uncertainty is one important barrier to plant investment. A second barrier is the perceived *technological risks* associated with building such large and complex facilities. A third barrier is the uncertainty over whether there will be a market for synthetic fuels in the future. These *market uncertainties* are due to the fluctuations of world oil supply and prices over the last decade, to differing projections of future supply and prices, and to concerns over whether the new fuel would be compatible with distribution systems and end-use technologies. The final barrier is simply *access to capital.* Synthetic fuels production plants are very capital-intensive and must compete with other projects for funds at interest rates which have recently been high.[3]

These barriers apply to different synthetic fuels in varying degrees. For methanol plant investment, technological risks are not as great as for other synthetic fuel processes, since both the gasification and synthesis processes have been commercially demonstrated, though at lower capacities than those projected for future coal-to-methanol plants.[4] While there would still be some concern over the economics of methanol production, it now appears that methanol is likely to be cost competitive with gasoline from imported oil by the 1990s, and the existence of a petroleum tax would strengthen that assertion. Access to capital would remain as a possible barrier, but capital flows to those projects that have the strongest probability of making maximum profit, so to some degree this barrier is an extension of the others. It appears that the single most critical barrier to methanol plant investment is the uncertainty over whether a market will exist for the methanol fuel. As noted above, this uncertainty has an economic component (will world oil prices be high enough that methanol is cost competitive with gasoline?) as well as an institutional component (will a distribution network and enough vehicles exist which are compatible with methanol fuel?). Obviously, large amounts of methanol fuel, no matter how inexpensive, are of little value if compatible end-use technologies do not exist or distribution is not feasible.

The optimum governmental financial incentive must therefore satisfy, as much as possible, two different sets of conditions. On the one hand, the incentive ought to minimize governmental intrusion and make maximum use of the marketplace incentive both to optimize economic efficiency and to allow a valid analysis of the cost performance of an individual project. On the other hand, the incentive must also convince investors that there will be a market for the methanol fuel at a price that allows a profit compared to projected production costs.

The major types of financial incentives can now be evaluated with respect to these criteria. Government loans and government-backed loan guarantees clearly fail to satisfy these criteria. With a loan guarantee backed by the federal government, a project sponsor faces a much lower cost of failure. Thus, much of the incentive for economic efficiency is lost. Loans and loan guarantees also hide the true cost of production, which would make it more difficult to assess the profit potential of additional plants. Finally, loans and loan guarantees do not ensure a market for the methanol product. The other ''front-end'' incentives (i.e., incentives which encourage investment rather than production)—investment tax credits, accelerated depreciation, and construction grants—also fail to resolve concerns about marketplace uncertainty. Furthermore, the former two are attractive only if the sponsor has considerable taxable income while the latter simply represents a large and inflexible governmental gift with little or no assurance of any return whatsoever. Production subsidies other than price guarantees, such as production tax credits or direct cash outlays based on production levels, also do not overcome marketplace uncertainties.

Of the major types of financial incentives, only price guarantees ensure that the investor will indeed have a market for methanol fuel. There are two primary types of price guarantees. Under a competitive fixed price guarantee, the private sector bids for a contract with the government, and the guaranteed price is set at the time of the contract. Under a nonnegotiated open-end price guarantee, the government agrees to guarantee a price at the time of sale which allows the sponsor a profit. Clearly the competitive fixed price guarantee maximizes economic efficiency by encouraging the owner to minimize costs at all times. Since the guaranteed price is fixed, any economic inefficiency is penalized. In addition, the fixed price guarantee minimizes governmental intrusion into the process since once the contract is signed, the government has no real role to play in plant design, financing, or operation. Competitive fixed price guarantees place maximum reliance on private decision making in the plant's planning, construction, and operation. Yet the government does assume the risks associated with marketplace uncertainties by guaranteeing a purchase price at the time the agreement is reached. A final important advantage of the competitive fixed price guarantee is that its value can be easily calculated, making overall evaluation of the economic feasibility of a methanol project fairly straightforward.

In summary, the federal government will have to play a role in developing a methanol fuel production industry. This role should be limited to one financial incentive that can be simply and fairly used to determine which projects to support, should minimize governmental intrusion and rely on private sector decision making as much as possible, and must nevertheless protect sponsors from serious marketplace uncertainties. Of all the financial incentives available to the government, the competitive fixed price guarantee is the only one which satisfies all of these criteria and should be the cornerstone of a federal program to encourage methanol plant investment.

### The Competitive Fixed Price Guarantee

We recommend the following general procedure for competitive fixed price guarantees for methanol production. The federal government would invite bids for a stated methanol price guarantee level over a specific time period. The government would accept those bids which conformed to the specified conditions up to a maximum production capacity on a first-come basis. Sponsors would have to show a commitment to construction and evidence of available capital. If sufficient production capacity had not been obtained in response to the first invitation, a second invitation would be initiated involving a slightly higher guaranteed price. This process would continue until either the desired production capacity was reached or the price guarantee level reached a maximum value. In any case, the final price guarantee level would be the same for all sponsors, equal to the level established in the final bid(s) accepted by the government, regardless of the level in the invitation to which individual sponsors originally responded.

Contracts would be signed between the government and the approved project sponsors for the agreed-upon price level, methanol volume, and time frame. When the contract becomes due, the government's involvement is enforceable only if the owner is actually producing methanol fuel. The owner sells the methanol in the marketplace at the market price. If the market price is greater than the contracted guaranteed price, then the government has no liability whatsoever. If the market price is less than the price guarantee, then the government must reimburse the producer for the difference between the guaranteed and market prices for the quantity of output up to the maximum volume in the contract. The government would always retain the option of direct purchase if the guaranteed price was greater than the market price. This process would continue throughout the time period agreed upon in the contract.

Several specific aspects of this proposal must be discussed further. One issue concerns the duration of the methanol price guarantees. Longer guarantees would generate greater interest on the part of methanol plant investors, but would also entail higher potential monetary risks to the federal government. We suggest that the price guarantees be available for five years after the agreed-upon production start-up date. This type of arrangement will encourage construction to be completed as early as possible, since delayed construction will reduce the period over which the price guarantee would apply.

The appropriate year to target the beginning of methanol fuel production under the price guarantee program would be 1990. Most analysts expect world oil prices to be somewhat stable through the rest of the 1980s but to begin to rise fairly significantly in the early 1990s. The latest long-term energy projections by the Department of Energy (DOE) support this supposition.[5] It is expected that coal-to-methanol plant construction could begin in 1986 or 1987. Assuming three to four years for construction, the initial plants could begin operation in 1990. This schedule for the initial large volumes of methanol fuel would be consistent with the methanol vehicle tax credit to be discussed in chapter 8, which would encourage the purchase of large numbers of pure methanol vehicles beginning in 1990. The most efficient transition would involve phasing-in of the methanol plants throughout each calendar year so that the total methanol fuel supply would be consistent with the fuel demands of methanol vehicles already on the road as well as those purchased during the year.

The initial and maximum price guarantee levels also need to be determined. Projections by the Environmental Protection Agency (EPA) and the Office of Technology Assessment (OTA) have shown that fuel methanol can be produced and sold at retail outlets for between \$10.00 and \$15.00 per million Btu (1982 dollars, excluding taxes).[6] This range translates to a cost of \$0.57 to \$0.86 per gallon of methanol. The price guarantee would be structured such that the methanol project sponsors would agree to provide the methanol to St. Louis, Pittsburgh, or Atlanta. Thus, the project sponsors would have to include long-range distribution costs but not local distribution or retail costs. Based on previous EPA work, local distribution and retail costs for methanol equal approximately \$0.07 per gallon.[7] Thus, the projected production cost of methanol (including long-range distribution cost) would be between \$0.50 and \$0.79 per gallon. Accordingly, we suggest that the initial bid invitation involve a guarantee of \$0.50 per gallon with successive invitations, if necessary, rising by \$0.05 per gallon increments. No invitation should be issued at a guarantee higher than \$0.80 per gallon. The invitations would be halted after the established capacity (to be discussed later) was achieved, and all the price guarantees would be at the level of the final successful bid. The actual guaranteed prices would be inflation-adjusted from the 1982 dollars basis used here.

We must also consider the issue of whether the government should impose any constraints on the feedstocks used in those methanol production plants eligible for price guarantees. One of the advantages of methanol is that there are many different feedstocks which can be used in its production: all types and grades of coal, natural gas, peat, wood, biomass, and even municipal waste. All of these resources are available in the United States and will likely be used in the future. However, a primary justification for this initial federal involvement in methanol production is to provide an alternative market for the high-sulfur coal currently used by eastern and midwestern utilities in order to encourage their conversion to low-sulfur coal as part of the overall strategy for resolving the acid rain problem. The methanol price guarantee program can serve to launch an alternative demand market for high-sulfur coal by requiring that eastern and midwestern high-sulfur coal be used as a feedstock for any methanol produced under the price guarantee. The higher coal sulfur levels are not a problem in methanol production, since sulfur is fairly easily removed during the gasification step of the process. Thus, we recommend that the methanol price guarantee be limited to only those projects which utilize eastern or midwestern coal with a sulfur content in excess of 1 percent as the methanol feedstock.

In order to help ensure that high-sulfur coal is utilized as a methanol feedstock in the early 1990s, thus contributing to the maintenance of the eastern and midwestern coal markets, we also recommend that consideration be given to a requirement that methanol produced from high-sulfur coal be granted an extension of its current exemption from the federal motor vehicle fuel tax. The Surface Transportation Assistance Act of 1982 exempted pure alcohol fuels produced from feedstocks other than petroleum and natural gas from federal fuel taxes until October 1, 1988.[8] We suggest that this exemption be continued from October 1, 1988, through October 1, 1993, for methanol made from high-sulfur coal. Methanol produced from other feedstocks would be taxed at the normal volumetric rate (presently, $0.09 per gallon). Beginning on October 1, 1993, we recommend that all methanol be taxed on an equivalent Btu basis with gasoline, e.g., if gasoline were still taxed at $0.09 per gallon, then methanol would be taxed at $0.045 per gallon.

Thus, this initiative would involve the federal government issuing an initial price guarantee invitation of $0.50 per gallon (1982 dollars) of fuel grade methanol delivered to St. Louis, Pittsburgh, or Atlanta. The price guarantee would apply for five years after the contractually agreed production start-up date and only for methanol produced from eastern and midwestern coal with a sulfur content of 1 percent or greater. The government would accept bids at $0.50 per gallon up to a maximum production capacity on a first-come basis. If there were an insufficient number of bids at $0.50 per gallon to reach the capacity ceiling, then the government would issue a second invitation with a guarantee of $0.55 per gallon. Successive invitations at levels of $0.05 per gallon higher would be issued until the capacity ceiling was reached, although no invitation would be issued with a guarantee in excess of $0.80 per gallon. The actual price guarantee for all projects would be equal to the guaranteed level involved in the final successful bid. The first year during which methanol would be produced under the price guarantee program would be 1990.

The final issue to be considered is the maximum production capacity covered by the price guarantee. A modest capacity ceiling would limit the financial risk and liability of the government, but might not be sufficient to maintain the demand for high-sulfur coal production. Two options for resolving this issue will be discussed. The first option is the more conservative approach, and it would stimulate the construction of only enough plants to supply sufficient volumes of methanol to accommodate the vehicles purchased under a one-year methanol vehicle tax credit program. This limited program would launch the methanol fuel industry by breaking down the initial implementation barriers and thus allow the marketplace to dictate the degree and pace of industry expansion. We would recommend 300,000 barrels per day (BPD) as an appropriate ceiling for this option. This would involve three 100,000 BPD methanol plants (if the entire outputs were covered by the guarantee) or a larger number of smaller plants.

The philosophy behind this option is that the 300,000 BPD methanol price guarantee would promote the first generation of large coal-to-methanol production plants and that the success of the initial plants would launch a large-scale methanol fuel industry. The 300,000 BPD price guarantee may not be sufficient, however. It will take approximately three to four years to complete construction of and begin methanol production in additional plants, and there could be a further delay in additional methanol production plant investments until the operating performance of the initial plants and the market performance of methanol fuel become clearly established. Yet in order to maintain a consistent market for high-sulfur eastern and midwestern coal, methanol fuel production must grow rapidly in the early 1990s. As discussed in chapter 2, analyses of various acid rain control programs which allow coal switching (as opposed to mandatory scrubbing) have indicated that between 90 million and 170 million tons of eastern and midwestern high-sulfur coal could be displaced. It has been projected that approximately 4.3 barrels of methanol can be obtained from each ton of bituminous coal in a large coal-to-methanol plant.[9] Thus, in order to maintain demand for, say, 130 million tons of high-sulfur bituminous coal, methanol demand would have to total approximately 560 million barrels per year or 1.5 million barrels per day (MBPD).

In order to ensure that the eastern and midwestern high-sulfur coal markets were maintained, a second option for the federal government would be to offer price guarantees which would ultimately total 1.5 MBPD. In order to permit an orderly phasing-in and to avoid construction bottlenecks, it would be advisable to spread out the price guarantees in five 300,000 BPD increments over five successive years. (The schedule will supply enough methanol so that about 50 percent of each year's new vehicle production could be methanol-fueled.) In other words, the federal government would offer the guarantees for methanol production which would begin in 1990, 1991, 1992, 1993, and 1994, respectively. Each year the mechanics of the price guarantee program would be similar to the process outlined. Each year's invitation and bidding process would be completely independent of previous years' agreements. The guaranteed price levels for each year's program would be equal to the highest accepted bid for that year, but there would be no relationship among the guaranteed levels of various years. A project accepted for one year could not be eligible for any future invitation.

This second option would ensure maximum stimulus for maintaining the high-sulfur eastern and midwestern coal markets. Of course, it does so by involving the federal government in the developing methanol fuel industry to a much greater degree, and entails the possibility of a much greater financial commitment. However, even with full implementation, the methanol covered by the price guarantee would be only a fraction of total U.S. highway fuel demand. DOE projects that 1995 gasoline consumption will be 4.5 MBPD and diesel consumption will be 1.6 MBPD.[10] Assuming that methanol is 20 percent more efficient than gasoline and similar in efficiency to diesel fuel, the 1.5 MBPD methanol price guarantee program would displace 20 percent of 1995 gasoline consumption or 13 percent of all 1995 highway fuel consumption.

The objective of the methanol price guarantee program is to convince both vehicle producers and operators that there will be an adequate and stable methanol fuel supply. By coordinating the timing of this program with other initiatives in the methanol implementation program, we can promote an efficient introduction of methanol vehicles and fuel. In addition to ensuring an adequate methanol supply, a price guarantee program could also have major impacts on employment and the federal budget.

### Increased Domestic Employment

The construction and operation of domestic coal-to-methanol plants would increase employment levels, particularly in high-sulfur coal regions such as the Midwest and northern Appalachia. We have considered two capacities for the methanol price guarantee program: 1.5 MBPD and 0.3 MBPD. The projected employment associated with a 1.5 MBPD domestic high-sulfur coal-to-methanol fuel production industry can be found in chapter 4 (see table 7 and accompanying discussion). Total employment would peak at approximately 400,000 jobs in 1989, and 44,000 permanent coal mine mouth jobs would be sustained by the 1.5 MBPD coal-to-methanol industry. Projected employment levels for the 0.3 MBPD price guarantee program would be approximately one-fifth of those of the 1.5 MBPD program.

### Possible Impacts on the Federal Budget

The actual costs of methanol price guarantees to the federal government are dependent on a number of factors which cannot be predicted in advance. Some of these factors would be fixed at the time the agreements were made, such as the guaranteed price level and the output quantities. One very critical factor, the market price for the methanol produced, would not be known until the methanol was sold, years after the contract was signed.

It is possible to calculate the cost to the treasury under various scenarios. The following discussion will be in terms of 1982 dollars and will exclude taxes. It has been stated that it appears that methanol could be produced and sold at retail outlets for between $0.57 and $0.86 per gallon. This figure includes all distribution costs from the production facility to the retail outlet, as well as retail markup. The equivalent delivered cost to major cities, which would exclude local distribution and retail costs, would be $0.07 per gallon less, or between $0.50 and $0.79 per gallon. To be conservative, we will assume that the government contracts for all of the methanol at a guaranteed price level of $0.80 per gallon, which would be the maximum price guarantee level allowed in the program.

The favorable case would be that the delivered marketplace value of methanol exceeded $0.80 per gallon (or, alternatively, if the retail pump value exceeded $0.87 per gallon, excluding taxes) for the five-year production time period. If so, then the methanol producers would sell their output on the market to the highest bidder and the government would have no liability whatsoever. The probability of this is actually rather high, in light of the probability of rising oil prices and the imposition of a petroleum tax. For example, assuming that methanol will be 20 percent more efficient than gasoline on a Btu basis, methanol will have a value of 60 percent of the value of gasoline on a per gallon basis. Accordingly, methanol at $0.87 per gallon would have equivalent value to gasoline at $1.48 per gallon or crude oil at $44 per barrel. $44 per barrel is not much above the peak world oil price in 1981 and is well within the range of 1995 world oil prices given in the most recent DOE projections, even if no supply disruption is assumed.[11] Such a disruption, or a petroleum tax, would make these oil prices much more probable resulting in a low likelihood of government liability for the methanol price guarantee.

A reasonable worst case scenario would involve assumptions of very low world oil prices ($30 per barrel), no petroleum tax, and that methanol would have a retail value of one-half that of gasoline on a per gallon basis (this ignores efficiency advantages of methanol use in spark-ignited engines). Crude oil at $30 per barrel results in retail gasoline of $1.00 per gallon, excluding taxes. Methanol would then have a retail value of $0.50 per gallon. With a maximum guaranteed price of $0.80 per gallon and an additional $0.07 per gallon cost to distribute and retail the methanol, the government would have to provide $0.37 per gallon to bring the methanol to market. For the one-time 300,000 BPD, five-year guarantee, this would result in a subsidy of (300,000 barrels/day) × (365 days/year) × (42 gallons/barrel) × ($0.37/gallon) = $1.7 billion per year or $8.5 billion over the five-year period during which the guarantee would apply. This would be a relatively small price to pay if the methanol price guarantee program contributed to stabilized world oil prices and resulted in an environmentally sound use for high-sulfur coal. If the larger 1.5 MBPD price guarantee option were implemented under these same assumptions, with the contracted production increasing by 300,000 BPD each year beginning in 1990, then the cost to the federal government would be $1.7 billion in 1990, increasing by $1.7 billion per year until it reached a maximum of $8.5 billion in 1994, then decreasing by $1.7 billion per year until it was zero by 1999. These outlays would total $43 billion over the nine-year period. While large in total expense, the outlays per year are not excessive, especially since they would occur only if world oil prices remain low for the next fifteen years, which would result in large U.S. oil import savings.

A conservative scenario would involve assumptions of $40 per barrel crude oil and a value of methanol equal to 60 percent of the value of gasoline on a volumetric basis. These assumptions would result in retail values of $1.34 per gallon for gasoline and $0.80 per gallon for methanol. With a maximum government outlay of $0.87 per gallon to bring the methanol to market, the total liability would be $0.07 per gallon. For the single 300,000 BPD price guarantee option, the cost would be $320 million per year or $1.6 billion over the five-year time period. For the 1.5 MBPD price guarantee program, the cost would be $320 million in 1990, rising by $320 million each year until it reached a maximum of $1.6 billion in 1994, and then decreasing by $320 million per year until it became zero in 1999. The total aggregate cost would be $8.1 billion over the nine years.

Table 22 summarizes the preceding analyses for the 1.5 MBPD price

TABLE 22. Projected Annual Cost to the Federal Government of 1.5 MBPD Methanol Price Guarantee Program under Various Scenarios ($ Billion)

| Year | Favorable Scenario | Conservative Scenario | Worst Case Scenario |
|---|---|---|---|
| 1990 | 0 | 0.32 | 1.7 |
| 1991 | 0 | 0.64 | 3.4 |
| 1992 | 0 | 0.96 | 5.1 |
| 1993 | 0 | 1.3 | 6.8 |
| 1994 | 0 | 1.6 | 8.5 |
| 1995 | 0 | 1.3 | 6.8 |
| 1996 | 0 | 0.96 | 5.1 |
| 1997 | 0 | 0.64 | 3.4 |
| 1998 | 0 | 0.32 | 1.7 |
| 1999 | 0 | 0 | 0 |
| Total cost | 0 | 8.1 | 43 |

guarantee proposal. The annual cost for the 1.5 MBPD price guarantee could range from zero to a maximum one-year cost of $8.5 billion. The conservative projection involved a maximum one-year outlay of $1.6 billion with an aggregate cost of $8.1 billion over nine years. Clearly, the accompanying highway fuel tax proposal (even without the expected higher world oil prices) will make it very likely that the government cost of this proposal is zero.

Finally, the extension of the federal fuel tax exemption from 1988 through 1993 for methanol produced from high-sulfur coal would reduce federal revenues slightly. Assuming implementation of the 1.5 MBPD methanol price guarantee program, and assuming that in absence of the exemption methanol fuel would be taxed at $0.045 per gallon (i.e., on an energy equivalent basis with gasoline), the exemption would reduce federal revenues by $0.2 billion in 1990, $0.4 billion in 1991, $0.6 billion in 1992, and $0.8 billion in 1993, after which the exemption would expire.

# CHAPTER 7 Methanol Requirement for Service Stations

Even if large quantities of methanol fuel and methanol vehicles are ready to be made available at competitive costs, the methanol implementation program will fail if service stations do not provide the critical link between fuel and vehicles at the appropriate time.

It can be argued that if both methanol fuel and the vehicles which require methanol are available and cost-competitive, then the marketplace will exert sufficient pressure on service station operators to offer methanol fueling. But the *timing* of the decision of the service station industry to offer methanol is absolutely critical to the overall success of the implementation program. If the vast majority of service station operators wait just one additional year to install methanol fueling capability, then there may well be no market for the initial large quantities of methanol fuel and the entire program might collapse. Because of the interdependence among fuel supply, demand, and distribution, it is imperative that all other parties involved have some assurance that methanol fueling stations will be available when needed.

The U.S. service station industry is large, diverse, and difficult to characterize. The Census Bureau counts service stations every five years, and estimates annual changes for other years. The last publicly available census data were taken in 1977. At that time, the census reported that there were 176,465 service stations in the United States.[1] The industry has been shrinking in the last few years, however, and the Commerce Department has estimated that there were 144,690 service stations operating in 1982.[2] The Census Bureau defines a gasoline service station as a retail business whose petroleum sales are 50 percent or more of its total dollar sales. Thus, the census data excludes many businesses which sell fuel, such as automotive garages, motels, car washes, convenience food stores, etc. A recent Lundberg Survey indicates that this undercount is very significant. It reported that there were actually 263,348 gasoline outlets in the United States in 1977, and 210,875 in 1982.[3] These data are 49 and 46 percent greater, respectively, than the 1977 and 1982 census figures.

There are several possible strategies that could be adopted by government to increase the likelihood of service stations offering methanol fuel. One would be a cost-sharing assistance program available to service station owners who voluntarily install methanol fueling capability. For example, the federal government might agree to reimburse service station owners for 50 percent of the cost of adding methanol fueling capability to an existing service station. The California Energy Commission has estimated that the maximum expense of adding methanol fueling capability, utilizing no existing facilities, would be approximately $28,000.[4] According to the Lundberg figures, there are currently approximately 211,000 service stations in the United States. If 75 percent of these stations took advantage of a 50 percent cost sharing program, and if on average the installation costs were $28,000, the government would have to provide $2.2 billion.

This relatively high cost is one disadvantage of the cost-sharing approach, especially since the addition of methanol storage and pumping hardware would ultimately occur anyway as methanol became more and more accepted. The cost-sharing program could be attacked as an unnecessary subsidy to the service station industry. Many stations should be able to utilize existing hardware (storage tanks, pumps) and thereby reduce the overall cost of conversion, but under a cost-sharing agreement the incentive to reduce overall costs might be overwhelmed by the desire of the owner to have new facilities subsidized by the government. A final drawback of this approach is that while it is a strong inducement for service station owners to add methanol fueling capability, it does not guarantee that it will occur.

A second strategy to encourage methanol retailing would be for the federal government to enter into methanol purchase guarantees at the retail level, i.e., the government would agree to purchase (or subsidize on a per gallon basis) whatever methanol the retailer did not sell to private customers up to an agreed amount. This could be a rather low-cost incentive to the government if methanol demand was at or in excess of expectations, but could be very expensive if projections were too high. A primary disadvantage of this incentive is the complexity of overseeing the actual sales of thousands of individual service stations. The administration of such a program could be very time-consuming and expensive.

A third strategy would be to allow tax credits for service station owners who add methanol fueling capability. This strategy is preferable to the previous two, in that it would probably be less expensive and more easily administered and enforced through existing Internal Revenue Service mechanisms. Still, tax credits would not, unless they were very high (in which case they approach direct subsidies), ensure that a sufficient number of service stations would invest in methanol storage and pumping hardware.

Another option is simply for the federal government to require certain large service stations to provide methanol retailing. Doing so would ensure the installation of methanol fueling facilities at the right time, which is necessary for the overall success of the implementation program. In addition, this goal would be achieved at no cost to the federal government. Of course, a governmental mandate would impose financial requirements on service station owners, but the regulations could be designed to minimize the impacts on small, low-volume service stations, which would be most vulnerable to new investment requirements.

There is a precedent for the federal government requiring certain service stations to offer a new transportation fuel. In 1973 automotive manufacturers announced that many 1975 model year production vehicles would utilize catalytic converters to reduce exhaust emissions to meet standards set by Congress and EPA. It was well known that the use of leaded gasoline would contaminate the catalysts and eventually render them ineffective. Catalyst-equipped vehicles would have to use unleaded gasoline in order to maintain compliance with emission standards, and EPA published regulations prohibiting the use of leaded gasoline in catalyst-equipped vehicles. Still, even with an assured demand for unleaded gasoline, there was considerable concern that many service stations might not respond quickly enough to prevent unleaded gasoline availability difficulties for purchasers of 1975 model year vehicles. Accordingly, EPA also promulgated regulations to ensure the general availability of unleaded gasoline compatible with the 1975 catalytic emission control systems.[5] These regulations, promulgated under Section 211 of the Clean Air Act which gives EPA authority to regulate the sale of fuels and fuel additives, originally established requirements at two levels. First, they provided that after July 1, 1974, every owner or operator of a retail outlet at which 200,000 or more gallons of gasoline had been sold during any recent calendar year had to offer for sale at least one grade of unleaded gasoline. Second, they provided that every owner or operator of six or more retail outlets had to offer for sale at least one grade of unleaded gasoline at no fewer than 60 percent of those outlets after July 1, 1974. In early 1974, EPA withdrew the second of these requirements, but expanded the unleaded gasoline availability requirement in rural areas to include those stations that sold in excess of 150,000 gallons of fuel per year.[6]

Several oil companies and refiners sought legal relief from the EPA regulations. The various suits were consolidated into one case heard by the U.S. Court of Appeals for the District of Columbia Circuit. In May 1974, the court upheld the regulations, agreeing with the petitioners only on a fairly minor point concerning liability. Most important, the court upheld EPA's "affirmative marketing requirement" of unleaded gasoline in light of Clean Air Act Section 211 and the great uncertainty that would otherwise exist over the availability of unleaded gasoline to fuel 1975 model year vehicles.[7]

Despite the concerns expressed by many in the gasoline refining and marketing industries, the transition to unleaded gasoline proceeded smoothly. It is estimated that approximately 120,000 (or 55 percent) of the 220,000 retail outlets nationwide were affected by the regulations and provided unleaded gasoline. Outlets unaffected by the regulations were primarily low-volume stations. While they would ultimately also choose to offer unleaded gasoline, the regulations allowed them to do so at their own pace. One major concern was that high-volume nonbranded independent retail outlets might have some trouble securing adequate unleaded gasoline supplies from the major petroleum refiners. This did not turn out to be a serious problem. In short, the federal requirement to provide unleaded gasoline retailing resulted in an efficient and timely transition for the service station industry.

A final option would be for the federal government to seek a voluntary commitment from the service station industry to provide methanol retailing. The advantages of such a strategy would be that it would involve neither major federal cost nor a federal mandate. The obvious drawback is that there would be no assurance that large numbers of retail outlet owners and operators would join such a program. It is true that approximately 80 percent of all service stations are owned by or have their fuel supplied by major oil companies.[8] The parent oil companies would probably be more receptive to methanol retailing than would independent retailers, and might even help finance necessary equipment changes. But again there is no guarantee, and probably a low likelihood, that service stations would react quickly enough to provide the necessary capacity at the beginning of large-scale methanol production.

In summary, because of the large capital investments required to produce methanol fuel and vehicles, ensuring that a sufficient number of retail outlets will offer methanol fuel is a critical part of any methanol implementation program. Although the marketplace would ultimately respond to the need for methanol retailing, precise timing is necessary to coincide with the production of large quantities of methanol fuel and the introduction of large numbers of methanol vehicles. Several strategies are available to the government to try to ensure that service stations offer methanol fuel, but only a federal mandate to require methanol fuel at certain retail outlets achieves the desired goal with maximum certainty and minimum cost. The precedent of the 1973–74 EPA regulations which effectively integrated unleaded gasoline into the national service station network is evidence that a similar requirement for methanol could be successful.

### High Volume Service Station Requirement

Legislation should be enacted that would explicitly authorize EPA or another federal agency to promulgate regulations which would provide for the general availability of methanol fuel by requiring certain retail outlets to offer it for sale. Congress itself could identify the framework and timing of the requirement, or the responsibility for doing so could be delegated to the appropriate federal agency. Regardless, the timing of the methanol retailing requirement must be consistent with other initiatives in the methanol implementation program which affect methanol fuel and vehicle production.

We propose that every owner or operator of a retail outlet in the United States which had sold 300,000 gallons or more of gasoline during 1985 or any succeeding calendar year be required to provide methanol fuel for sale by January 1, 1990. 1990 is the appropriate year for this requirement since that is when both the methanol price guarantee and methanol vehicle tax credit programs would begin, creating both large-scale methanol supply and demand. The fuel itself would have to be in conformity with methanol fuel specification regulations published by EPA and used by automotive manufacturers for emission and fuel economy certification.

There will be two major impacts of this proposal to require certain retail outlets to offer methanol fuel for sale—an increase in the general availability of methanol fuel and an economic impact on those service stations which will have to add methanol fueling capability.

### Availability of Methanol Fuel

In view of the ever-changing nature of the service station industry, it is impossible to predict the exact number of stations that would be required to install methanol fueling capability by 1990, and the percentage of total U.S. fuel which the affected stations distribute. Nevertheless, some projections can be made. Data published by the Census Bureau and the Lundberg Letter have been analyzed to characterize the service station industry by annual petroleum throughput.[9] The results are summarized in table 23. It can be seen that approximately 93,000 service stations sell over 300,000 gallons of fuel annually. This is only 44 percent of all retail gasoline outlets. But table 23 shows that these high-volume stations account for 77 percent of all fuel sold on the retail market. A requirement that these high-volume stations market methanol fuel would ensure wide availability of methanol throughout the United States. It is likely that once the methanol implementation program

TABLE 23. Characterization of the U.S. Service Station Industry by Annual Petroleum Throughput Range (Thousands of Gallons per Year)

| | 0–120 | 120–300 | 300–600 | 600–1,200 | 1,200+ | Total |
|---|---|---|---|---|---|---|
| Number of stations | 54,800 | 63,300 | 55,900 | 29,500 | 7,400 | 210,900 |
| Percentage of total stations | 26 | 30 | 26.5 | 14 | 3.5 | 100 |
| Percentage of total throughput | 4 | 19 | 29 | 28 | 20 | 100 |

accelerated even the smaller retail outlets would add methanol fueling capability.

It must be noted that part of the data base utilized in generating the results shown in table 23 date back to 1977. Given this fact, and the continuing trend in the service station industry to smaller numbers of higher-volume stations, it is likely that the industry of the 1990s will be significantly different than the one characterized in table 23.

### Aggregate Cost to the Service Station Industry

Service station owners and operators will have to spend additional capital resources to accommodate methanol sales. The extent of the necessary investments depends upon the ability of owners to utilize existing hardware to minimize additional costs. We will define a range of costs involving worst case and best case assumptions.

The worst case situation would be if all of the affected retail outlets had to purchase new equipment and were not able to utilize any existing facilities. Table 24 gives a breakdown of the cost to a service station owner of adding a completely new methanol fuel facility.[10] This table is based on actual costs which the California Energy Commission has incurred in adding methanol fueling capability at eighteen different sites (additional stations are planned). The total is $28,000 per site. Table 23 showed that this proposal would affect approximately 93,000 retail outlets. If in fact each of these outlets had to spend $28,000 to add an entirely new fueling unit, the total cost to the industry would be $2.6 billion.

It is likely that many stations will not have to add a completely new methanol fueling configuration, however. As can be seen in table 24, one-half

TABLE 24. Projected Cost of Adding Methanol Fueling Capability at an Existing Site Utilizing All New Hardware

| Item | Cost |
|---|---|
| 10,000-gallon carbon steel tank | $2,600 |
| Pump | 1,200 |
| Dispenser | 1,000 |
| Other hardware | 5,200 |
| Excavation/installation of tank | 11,000 |
| Contractor fee | 1,000 |
| Other contractor expenses | 5,000 |
| Miscellaneous | 1,000 |
| Total | 28,000 |

of the cost of adding methanol fueling equipment is due to the purchase and installation of the storage tank. A large part of the contractor services would also likely be associated with excavation and installation of the storage tank. Approximately 90 percent of the storage tanks now in use are made of carbon steel, which is generally satisfactory for methanol storage (the other 10 percent are fiberglass tanks; some of the standard resins used in fiberglass gasoline tanks are generally considered unacceptable for methanol storage).[11] Thus, if a service station could free up one of these tanks from a different fuel it could be used for methanol storage. Fortunately, this seems to be a very realistic option. Most current service stations have both leaded and unleaded gasoline storage tanks (many stations also have tanks for diesel fuel). Most stations have two brands of unleaded gasoline, regular and premium, requiring separate tanks. By 1990, it is anticipated that there will be very few noncatalyst passenger vehicles on the road. There will still be noncatalyst gasoline-fueled trucks, but these trucks are typically in fleets and often have private fueling facilities. Thus there will be very little demand for leaded gasoline.

Retail outlets would be able to use the carbon steel tanks which had been used to store leaded gasoline to store methanol. The only possible increased cost for the tank would be for cleaning or water protection. It is also possible that existing pumps, piping, and dispensers may be modifiable for methanol use. To be conservative, however, we will assume that new pumps, piping, dispensers, and other hardware will be installed due to materials compatibility problems, higher pumping rates, etc. Assuming that no new tanks would be needed but that new pumps, piping, and dispensers will be necessary, and that only a small contractor fee will be necessary, we project that each outlet affected by the regulations would have to spend approximately $9,500. This would total $880 million for the 93,000 stations estimated to be affected. It should be noted that it may be possible to utilize existing facilities even more, and that part of the cost of any new hardware could be considered to be modernizing cost which would be necessary at some time in the future anyway. Thus, the real additional cost could be less than the total expenditures.

# CHAPTER 8 Federal Tax Credit for Methanol Vehicle Purchases

Strategies such as a methanol price guarantee and a petroleum tax should ensure that significant methanol fuel supplies will develop by the early 1990s. Though the initial (first-year) methanol capacity, 300,000 barrels per day, would be only a small fraction of that which will ultimately develop, it will nevertheless represent an amount of methanol that will dwarf the nontransportation demand for methanol by the chemical industry (current U.S. methanol production is approximately 100,000 barrels per day).[1] Thus, it is imperative that a sufficient transportation demand for methanol exist by the early 1990s or the initial coal-to-methanol production will be underutilized, and price guarantee supports would be necessary.

The most important figure in the demand for methanol fuel is the individual vehicle purchaser. Automotive manufacturers have repeatedly stated that they are ready and willing to produce methanol vehicles when customer demand warrants. But while the entire nation will benefit from methanol usage in the long term, there are several factors inhibiting most individual vehicle purchasers from considering a methanol vehicle in the initial years of the program.

First of all, there is the inertia regarding change that is typical for most consumers, the "fear of the unknown." A healthy skepticism of major technological novelties is often appropriate and beneficial, but in this case many citizens will probably overestimate the degree of the change and the risks involved in owning a methanol vehicle. Methanol was one of the earliest motor vehicle fuels and is an excellent fuel for spark-ignited Otto-cycle engines. Petroleum fuels displaced methanol not because they were inherently better fuels, but because they became much cheaper after the discovery of large domestic oil fields. The racing industry today utilizes methanol fuel. Modifications required to produce methanol vehicles will likely be very straightforward and will not involve any fundamental design changes. Nevertheless, because vehicle fueling is one of the few maintenance tasks that most vehicle owners perform, many people associate vehicle type with fuel type and would consider a methanol vehicle to be significantly different from a gasoline vehicle. Many people would understandably avoid what they might perceive as a "new technology" when spending several thousand dollars on a new vehicle purchase. Related to the fear of technological change is the tendency of many consumers to consciously avoid a new vehicle model for a year or two in order to let other vehicle owners be the "guinea pigs." This tendency has been noted particularly with respect to automobiles, and is justified by data showing that it often takes a year or two of field experience to "work out the bugs" of a new model design.

Economic factors may also discourage individual consumers from initially purchasing methanol vehicles. In the initial years it is possible that methanol fuel may be slightly more expensive than gasoline, depending on world oil prices and the level of petroleum tax. There may also be some initial uncertainty over the future supply of methanol, since a vehicle owner must depend on supply of a fuel for several years into the future. The implementation of a price guarantee program and the successful operation of domestic coal-to-methanol plants should allay this concern, but it could remain as a possible constraint on methanol vehicle demand, especially during the first couple of years of vehicle availability.

Left entirely to the marketplace, methanol vehicles at first would probably appeal only to those who have a penchant for innovation. Initial vehicle sales would likely be very low and would grow slowly each year as more and more consumers become familiar with the individual and national benefits of owning and operating methanol vehicles.

There are several possible financial incentives which the government could adopt to spur methanol vehicle demand. The government could increase the likelihood of the availability of methanol vehicles by granting auto manufacturers incentives such as direct grants, low-interest loans, production tax credits, or accelerated depreciation benefits, all tied to methanol vehicle investment or production. Alternatively, the government could encourage private consumers to purchase methanol vehicles by offering low-interest loans or federal tax credits for the purchase of methanol vehicles, or federal tax credits for methanol fuel consumption.

There are several conditions that a financial incentive should satisfy in order to successfully stimulate methanol fuel demand. First, the incentive ought to be specifically targeted at the most important link in the methanol vehicle demand process—the decision by individual consumers whether to buy a methanol vehicle. This means that the incentive ought to be targeted at the *consumer* and not at the vehicle manufacturer; it is very possible that incentives directed at the manufacturers might encourage methanol vehicle production but not result in customer interest. On the other hand, if consumers demand methanol vehicles manufacturers would quickly provide them. Further, this means that the incentive ought to specifically reduce the purchase price of the vehicle, as opposed to reducing other costs of owning and operating a methanol vehicle. Second, the incentive should be structured to be significant enough at first to overcome the constraints discussed above and encourage sales in the first year of the program, and yet be phased out quickly so that the marketplace can guide future decisions. Third, the incentive must be simple for the general public to understand and easy for the government to administer. Federal tax credits for methanol vehicle purchases fulfill these conditions much better than any of the alternative governmental incentives. Tax credits reduce the purchase price to the consumer, can be quickly phased out, are easily understood by the taxpayer, and can be administered and enforced by the Internal Revenue Service just as are other tax credits.

In summary, the implementation of a methanol program requires assurance of significant methanol vehicle fuel demand. The most important figure for methanol demand is the individual vehicle customer, but there are several reasons why consumers might be hesitant to purchase methanol vehicles when they first become available. A financial incentive is necessary to ensure that consumers will buy a substantial number of methanol vehicles, and a federal tax credit is the simplest and most appropriate incentive to accomplish this goal.

## Design of the Tax Credit

We recommend the establishment of a federal tax credit for methanol vehicle purchases. The tax credit would apply to any person or corporation buying a new passenger car or light-duty truck (less than 8,500 pounds gross vehicle weight rating) that was certified by the manufacturer for use with a fuel containing at least 85 percent methanol.

We propose consideration of two options for the magnitude and duration of the methanol vehicle tax credit. One option would involve a $500 per vehicle tax credit for methanol vehicles purchased in 1990 up to a maximum of 5 million vehicles. This limited one-year program might be sufficient to ensure significant consumer interest in methanol vehicles if other parameters, such as methanol vehicle sticker prices, world oil prices, and methanol fuel costs, are favorable by 1991. These variables, which cannot be accurately projected at this time, may not be attractive enough, however, and a more comprehensive vehicle tax credit program would probably be necessary to ensure sufficient methanol fuel demand. Thus, a second option would involve tax credits of $1,000 for methanol vehicles purchased in 1990, $800 for methanol vehicles bought in 1991, and $600 for methanol vehicles purchased in 1992. Again, the credits would only apply to the first 5 million methanol vehicles sold in each calendar year. This multiyear program would be more expensive to the federal treasury but would provide greater certainty for methanol fuel producers and vehicle manufacturers. The tax credit would apply only to citizens purchasing new vehicles from authorized dealerships and maintaining ownership of the vehicles for a minimum of one year, and the credit could not be transferred from one taxpayer to another.

The levels of the tax credit could be adjusted upward or downward, of course, depending on the outlook for methanol vehicle sales at the time of congressional passage. Once the structure of the credit is set, however, it should not be changed, since the certainty of the tax credit is an important factor in its success. The credit needs to be large at the outset of the methanol implementation program when fuel demand is necessary but consumer reluctance is highest. The incentive would be quickly phased out, as it should become less necessary as consumer inhibitions recede. Tax credits are often expressed as a percentage of total outlays, up to a maximum level, but we feel that a fixed credit is more appropriate in this instance. With the fixed amount, the lower the methanol vehicle price the greater the incentive. For example, a $1,000 credit would be a 17 percent reduction in price for a $6,000 car but only a 10 percent reduction for a $10,000 car. Thus, the fixed credit encourages vehicle manufacturers to offer low prices for methanol-fueled models.

The timing of the tax credit is critical, and should not be changed once it is set. Methanol fuel and vehicle producers would see the tax credit as ensuring fuel and vehicle demand beginning in a certain year. Based on the proposed price guarantee program, the first large U.S. coal-to-methanol plants would begin producing fuel in 1990, with capacity increasing throughout the early 1990s. This production schedule would be consistent with beginning the vehicle tax credit in 1990.

The proposed federal tax credit would apply to the purchase of any new methanol-fueled passenger car or light truck regardless of whether or not it was built in the United States. Perhaps this initiative should also be designed to stimulate the development of a domestic methanol vehicle industry. It is reasonable to ask whether a U.S. tax credit ought to apply to the purchase of an imported methanol vehicle.

The primary impact of limiting the methanol vehicle tax credit to domestic vehicles would be larger sales of domestic methanol vehicles. This would result in greater employment in the U.S. auto sector and related industries and retain American dollars to stimulate economic growth. Our balance of payments would improve. It might well encourage foreign automotive manufacturers to invest in U.S. methanol vehicle manufacturing capacity in order to take advantage of the tax credit and the growing methanol vehicle market. The primary drawback of limiting the tax credit to domestic vehicles is the anticompetitive impact. Consumers would have fewer vehicles to choose from if they desired to qualify for the tax credit and manufacturers without U.S. plant capacity would be at a competitive disadvantage with respect to pricing policies. Thus, a decision to limit the applicability of the tax credit to domestic vehicles involves a balancing of the macroeconomic advantages with the anticompetitive effects.

The federal tax credit for methanol vehicle purchases could have major impacts in three areas: methanol vehicle sales and fuel demand, the federal budget, and the federal tax system. These impacts are discussed in the following sections.

## Increased Methanol Vehicle Sales and Fuel Demand

The goal of the federal tax credit is to ensure a much more rapid and managed introduction of methanol vehicles into the national transportation fleet than might otherwise be the case, thereby increasing methanol fuel demand in a coordinated manner. We anticipate that consumers would utilize the tax credit to the fullest, purchasing 5 million vehicles in 1990 under the one-year program and 5 million methanol-fueled vehicles per year in 1990, 1991, and 1992 under the multiyear program. Based on projected average sales of 10 million passenger cars and light trucks per year, sales of methanol vehicles would constitute approximately 50 percent of all new passenger vehicle sales in those years.

To anyone familiar with the automotive industry these projected sales volumes for a new type of vehicle might seem very optimistic. But it must be emphasized that rarely are new models introduced with significantly *lower* prices than competitive models. In recent congressional testimony, representatives of two domestic manufacturers stated that methanol vehicles would likely cost no more to produce than gasoline vehicles for production volumes greater than 100,000 vehicles.[2] The large federal tax credit program should ensure that methanol vehicles will be priced considerably lower than their gasoline-powered counterparts. In addition, there will probably be considerable competition among manufacturers to take the lead in the initial methanol vehicle market.

Table 25 gives the projected methanol fuel demand of vehicles which would be purchased under the two methanol vehicle tax credit options outlined earlier. These projections assume that the 5 million methanol vehicle ceiling is reached during each year that the tax credit is available, that no other methanol vehicles are sold in those years, and that the average new methanol vehicle is driven 12,000 miles per year and achieves 15 miles per gallon of methanol (assuming that methanol vehicles are 20 percent more energy-efficient than new gasoline vehicles having an average on-road fuel economy of 26 miles per gallon). By the end of the one-year tax credit program, methanol demand would be 260,000 BPD. With the multiyear program methanol demand would grow to 520,000 BPD by the end of 1991 and 780,000

TABLE 25. Projected Impact of Federal Tax Credit Options on Vehicle Sales and Tax Expenditures

| Year | Tax Credit ($) | New Methanol Vehicle Sales (million) | Total Methanol Vehicles at Year End (million) | Methanol Demand at Year End (BPD) | Tax Expenditures ($ billion) |
|---|---|---|---|---|---|
| Option 1 | | | | | |
| 1990 | 500 | 5 | 5 | 260,000 | 2.5 |
| Option 2 | | | | | |
| 1990 | 1,000 | 5 | 5 | 260,000 | 5 |
| 1991 | 800 | 5 | 10 | 520,000 | 4 |
| 1992 | 600 | 5 | 15 | 780,000 | 3 |

BPD by the end of 1992. Under the multiyear tax credit program, there would be 15 million methanol vehicles in use by 1992. This would be a sufficient number to expose all consumers to the methanol alternative. It would be expected that the methanol option would be seen as an attractive alternative because of more powerful and efficient engines. It is also likely that gasoline prices will have risen, due to higher world oil prices and the petroleum tax, to such a degree that operation of methanol vehicles will be significantly less expensive than gasoline and diesel vehicles. Accordingly, we expect methanol vehicle sales to remain high after the expiration of the federal tax credit program. If methanol vehicle sales remain at 5 million vehicles per year after 1992, then methanol fuel demand would rise to over 1 MBPD by 1993 and to over 1.5 MBPD by 1995, which would guarantee continued demand for high-sulfur coal.

### Federal Tax Expenditures

Of course the federal tax credit would have a deleterious effect on the federal budget by decreasing the amount of tax revenues collected. Measures which reduce tax revenues are often referred to as tax expenditures. The maximum tax expenditures that would result from the tax credit program discussed above are also shown in table 25. For the one-year tax credit in 1990, the maximum tax expenditure would be $2.5 billion. Under the three-year program the maximum annual tax expenditures would be $5 billion in 1990, $4 billion in 1991, and $3 billion in 1992. The maximum aggregate tax expenditure for the multiyear program would be $12 billion.

This element of the methanol implementation program would directly stimulate methanol consumption and thus directly reduce gasoline consumption, resulting in lower oil imports. For example, assuming that 5 million vehicles are sold in each of the years during which the tax credit program exists, table 26 gives the number of methanol vehicles in use at the middle of each year, the gasoline that is displaced by the use of those methanol vehicles (assuming 26 miles per gallon of gasoline and 12,000 miles per year), and the reductions in the U.S. oil import bill assuming that all of the reductions would be reflected in oil imports. The value of crude was taken to be $30 per barrel to be consistent with the $0.40 per gallon highway fuel tax which would be expected to cap U.S. oil import prices at approximately $30 per barrel. The low values for oil import savings reflect the usage of $30 per barrel for gasoline as well, which is clearly low since gasoline is a premium part of the crude barrel. The high value reflects the use of a factor of 1.4 to account for this premium, and results in a value for gasoline of $42 per barrel. Table 26 shows that the United States would export $0.8–1.1 billion less in 1990 for imported oil if either tax credit program were adopted. Under the multiyear program, oil import savings would increase to $2.5–3.4 billion in 1991 and $4.1–5.8 billion in 1992. These savings result from the substitution of domestically produced methanol for imported crude oil.

An examination of tables 25 and 26 indicates that the federal tax credit expenditures would exceed the oil import savings each year the credit would be in effect, except for the third year of the multiyear program. But while the tax credit would apply for only one or three years, methanol vehicles purchased under the program would continue to be operated for several years thereafter, providing significant oil import savings. While the one-year tax credit would cost the federal government $2.5 billion in 1990, methanol vehicles purchased with the tax credit (assuming 100,000-mile lifetimes) would reduce oil import payments during the 1990s by $14 billion to $19 billion as shown in table 26. Methanol vehicles purchased during the three-year tax credit program would provide oil import savings of between $41 billion and $58 billion over their lifetimes, which is much larger than the $12

TABLE 26. Savings in Oil Import Payments Based on Projected Methanol Vehicle Sales

| Time Frame | New Methanol Vehicle Sales (million) | Total Methanol Vehicles at Mid-Year (million) | Gasoline Displaced (million barrels) | Savings in Oil Imports ($ billion) |
|---|---|---|---|---|
| Option 1 | | | | |
| 1990 | 5 | 2.5 | 27 | 0.8–1.1 |
| Lifetime | — | 5 | 458 | 14–19 |
| Option 2 | | | | |
| 1990 | 5 | 2.5 | 27 | 0.8–1.1 |
| 1991 | 5 | 7.5 | 82 | 2.5–3.4 |
| 1992 | 5 | 12.5 | 137 | 4.1–5.8 |
| Lifetime | — | 15 | 1,370 | 41–58 |

billion tax expenditure. In addition, the tax credits return money to U.S. citizens to spend as they wish and to turn over in the U.S. economy, while the savings in oil imports are likewise retained in the United States for investment and consumption. The federal tax credit appears to be an excellent investment for the federal government.

### Effects on the Federal Tax System

Administration and enforcement of the federal tax credit would be carried out by the Internal Revenue Service (IRS). The additional costs to IRS of administering the federal tax credit should be small since tax credits are a common tax mechanism. IRS would have to slightly modify the tax form but this is generally done each year for other tax law changes. The most efficient process for administration of the program would likely be for IRS to establish a specific phone number for auto dealers to call to register new methanol vehicles and to verify that the tax credit was still available. Enforcement by the IRS should also be quite straightforward. IRS should have little trouble identifying and verifying what constitutes a methanol vehicle purchase for tax purposes.

The impact on individual taxpayers should also be minimal. Of course, there will be no impact on those taxpayers who do not purchase methanol vehicles or who choose not to take advantage of the credit. For those claiming the tax credit, all that is necessary is to give the date of the methanol vehicle purchase, the amount of the tax credit, and possibly the identification number of the vehicle for which the credit is being claimed on his or her IRS tax form. It should be noted that tax credits apply to taxpayers regardless of whether they itemize deductions or not, so the federal tax credit would not force additional taxpayers to use the long tax forms.

In summary, the administration and enforcement of a federal tax credit for methanol vehicle purchases is about as simple and inexpensive as any new program could be. The net impacts would be very negligible on both taxpayers and the IRS.

CHAPTER 9

# Methanol Requirement for Urban Buses

Over one thousand urban areas in the United States have intracity bus transit systems. Our largest cities typically utilize rail transit as well, but the great majority of cities rely exclusively on the intracity bus for their transit needs. These transit systems include 62,100 buses (some of which are small vans or which are used only as backup buses), which traveled 1,668,000,000 miles, carried 5,670,000,000 passengers, and consumed 11,500,000 barrels of petroleum in 1982.[1] Urban buses consume approximately 31,500 barrels per day (BPD) of petroleum.

The new intracity bus market is rather small in absolute numbers—typically three to five thousand new urban buses are sold to U.S. transit systems each year.[2] Two U.S. bus coach manufacturers, General Motors Corporation Truck and Coach (GMC) and Flxible, have supplied most of the new urban bus market in the last few years with their advanced-design buses. Recently, however, the list of bus coach options has been expanding. Several transit authorities have purchased "new look" buses from GM-Canada and small U.S. companies that, despite the name, involve more traditional bus designs at a somewhat lower cost. At the other end of the market, several importers such as M.A.N. (West Germany) and Icarus (Hungary) have established assembly plants in the United States to produce articulated buses which can provide higher seating capacities for high-volume transit systems.

The federal government is a major financial supporter of bus transit systems. The Urban Mass Transportation Act of 1964, as amended, contains formula-based programs for both capital and operating assistance. The formula apportions general revenue funds to an urban area based on a combination of total population, population density, ridership, and other factors. The federal government, through the Urban Mass Transportation Administration (UMTA), provides up to 80 percent of the funds used for capital purchases and up to 50 percent of the operating assistance required by transit systems under the formula program. UMTA also administers a discretionary capital grants program which can provide up to 75 percent of the funds for acquiring or improving capital equipment and facilities. This discretionary program is funded by the Mass Transit Account of the Highway Trust Fund, which receives $0.01 per gallon of the federal motor fuel excise tax. In fiscal year 1983, UMTA provided approximately $1.7 billion to bus operators, with approximately half for capital purchases and half for operating assistance.[3]

Nearly all urban buses are powered by diesel fuel—96 percent use diesel, 3 percent use gasoline, and less than 1 percent use propane.[4] The diesel engine is used in urban buses because of its much higher efficiency during low-speed, stop-and-go driving compared to the gasoline engine. Both GMC and Flxible coaches have almost exclusively utilized engines from GMC's Detroit Diesel Allison (DDA) Division. For many years the standard bus engine was the DDA 71-series engine, typified by the DDA 6V-71, a naturally aspirated, two-stroke, six-cylinder diesel engine, and its eight-cylinder counterpart, the DDA 8V-71. The 71-series engines are installed in over 90 percent of the urban buses currently operating in the United States.[5] Recently, the 71-series engines have been replaced in most new bus applications by the DDA 6V-92TA, a turbocharged, two-stroke, six-cylinder diesel with somewhat lower fuel consumption and emissions. Several additional diesel engine manufacturers have entered, or are expected to enter, the bus engine market. Some of the new coach manufacturers, such as M.A.N., build their own bus engines. In addition, the Cummins Engine Company, a leader in heavy-duty diesel truck engines, has indicated an interest in the bus engine market.

There are two grades of distillate fuel oil used in bus engines: No. 1 diesel and No. 2 diesel. No. 2 diesel is a much more common fuel, used nearly universally in trucks, passenger cars, and locomotive engines. No. 1 diesel is a somewhat higher quality fuel, with a lower boiling point, lower sulfur, and a lower aromatic content. Because of its composition, No. 1 diesel fuel provides for somewhat better performance in engines which undergo wide variations in speed and load and/or which must operate in lower temperatures, and generally produces less smoke. Accordingly, urban buses are a primary market for No. 1 diesel, especially in cities with low ambient temperatures and/or cities concerned with smoke and particulate emissions. Because of its higher quality, No. 1 diesel typically costs between $0.05 and $0.10 per gallon more than No. 2 diesel. At this time, some transit authorities use No. 1 diesel, some use No. 2 diesel, and some use a combination of both, either in a blend or based on the time of year.

There are two significant problems in relying on diesel as a universal transit bus fuel. The first drawback is air pollution. While diesel combustion offers some environmental advantages compared to gasoline combustion, such as lower carbon monoxide and evaporative hydrocarbon emissions, the diesel engine inherently produces much higher levels of several pollutants, most notably particulate matter and oxides of nitrogen (NOx). Urban buses are responsible for a large portion of the total diesel particulate loading in some cities. Because transit buses operate on the most heavily populated urban roadway corridors, and emit pollution at ground level, public exposure to diesel bus pollution is much higher than that to most other sources on an equivalent mass basis. Two other characteristics of diesel combustion, odor (due to certain organic species which often adsorb onto particulate matter) and black smoke (the visible component of particulate matter), while not considered independent pollutants because of their relationship to particulate matter, are major irritants to urban dwellers.

No. 1 diesel is a higher quality fuel than No. 2 diesel and its use can ameliorate emissions problems to some degree. EPA testing has shown that the use of No. 1 diesel can significantly reduce smoke, and can reduce particulate, oxides of nitrogen, and carbon monoxide emissions by 10 to 20 percent. On the other hand, hydrocarbon emissions are increased 10 to 20 percent by the use of No. 1 diesel, and fuel consumption is 1 to 3 percent better with No. 2 diesel.[6] While the use of No. 1 diesel can reduce most emissions somewhat, engines using No. 1 diesel still produce large quantities of particulates, oxides of nitrogen, and smoke.

EPA and local environmental authorities probably receive more complaints about bus pollution than pollution from all other mobile sources combined. This should not be surprising, since most large diesel trucks frequently operate outside of urban areas and exhaust from gasoline-fueled cars is generally colorless and odorless (though, of course, not harmless). Transit buses are operated in urban areas, generally on very congested streets, and are typically accelerating or decelerating, either of which is likely to produce smoke. Congress and EPA have recognized the environmental problems of the diesel bus for years, and in March 1985, EPA promulgated more stringent standards for particulate and oxides of nitrogen emissions. Still, the new standards will not take effect until 1991 and are expected to involve both an increase in new engine price and a slight fuel economy penalty.[7] In general, the combustion of methanol produces little or no particulate matter or smoke and NOx levels are approximately one-half of those of diesel engines.[8] The opportunity to significantly improve the urban environment provides a compelling motivation to utilize methanol as a fuel for urban buses.

A second major problem involved with reliance on diesel fuel for transit buses is the possibility of supply constraints or disruptions. This issue is twofold. On the one hand, diesel fuel supplies in general have been tightening up. The primary reason for this has been the shift in relative gasoline and diesel demands, as diesel vehicles have become more popular in nearly all automotive applications. It is getting more difficult to produce the necessary amount of diesel fuel out of a barrel of crude oil, especially with lower gasoline demand. Evidence of this problem is apparent, with No. 2 diesel fuel retail prices generally exceeding regular leaded gasoline prices, and even exceeding regular unleaded gasoline prices in some cities (just a few years ago diesel fuel was considerably cheaper than leaded gasoline). With continued dieselization, it is expected that this pressure on diesel supplies will worsen. At minimum this will continue to result in upward price pressures, at worst it could cause occasional supply difficulties.

Even more critical than this gradual tightening of the diesel fuel market, however, is the ever-present possibility of a world crude oil supply disruption caused by political conflict or coercion. The United States still imports approximately one-third of its crude oil, and events of the last decade have proven the extent to which our oil dependence can be exploited. Nearly all diesel fuel is used for high-priority purposes, and it is not clear whether urban buses could be guaranteed sufficient diesel fuel supplies in a crisis situation. Utilizing a fuel such as methanol, which could be made from domestic energy feedstocks, would guarantee a secure and stable fuel supply for transit authorities. This is a second major reason to consider methanol as a transit fuel.

It can be similarly argued that methanol fueling would be preferable for nearly all motor vehicles. There are, however, a number of reasons why a methanol-fueled urban bus program could be implemented relatively easily as compared to a general fleet transition to methanol.

First, because most transit authorities are public agencies, they should be sensitive to public complaints about the environmental problems of diesel buses and the benefits to be gained from operating methanol buses. Transit agencies typically receive relatively large operating subsidies from local units of government, and this provides a leverage point for local citizens interested in reducing air pollution. Even more directly, since the federal government provides up to 80 percent of the funds used to purchase new urban buses, it could directly promote interest in methanol buses by providing financial inducements for technology transfer or by simply requiring that all federal monies be used for methanol bus purchases. Second, transit authorities have centralized fueling sites which could be modified to store and dispense methanol fairly easily. This means that a methanol bus implementation program would be largely immune from the distribution problems associated with the widespread transition to a fuel like methanol, with its different chemical properties, requiring that scores of thousands of private service stations be capable of storing and dispensing it. Finally, because transit systems also have centralized maintenance facilities, there would be far fewer concerns over the proper maintenance and repair of a "new" or at least different engine technology. Thus, not only is a methanol bus program a good idea, but it is an idea which could have a high likelihood of being smoothly and successfully implemented.

Methanol's potential as an urban bus fuel has attracted worldwide interest. Three manufacturers, M.A.N. (West Germany), Daimler-Benz, and

General Motors' Detroit Diesel Allison Division, have developed methanol-fueled engines for urban bus applications.[9] As of mid-1984, a total of seven methanol buses were operating in San Francisco, Berlin, Auckland, and Pretoria. Many of these buses had accumulated between 50,000 and 100,000 miles; while each of the buses had some minor problems, in general both the manufacturers and the demonstration sponsors have been very pleased with the performance of the methanol buses.

The most important methanol bus demonstration program for the United States, both because of the manufacturers involved and its accessibility, is the one in San Francisco sponsored by the California Energy Commission. Current funding is being used to operate two GM and M.A.N. methanol buses, which went into service at the Golden Gate Bridge, Highway, and Transportation District in January and July of 1984, respectively, until spring 1985, and additional funding is being sought to continue the demonstration. The program has been designed to provide important information about operating cost, fuel and oil consumption, emissions, maintenance, driveability, durability, and consumer and driver reaction.[10] In general, the demonstration has been very successful in proving the feasibility of methanol transit buses. The M.A.N. bus has been particularly impressive with very few maintenance problems and an energy efficiency equivalent to its diesel engine counterpart both in service and on the Society of Automotive Engineers bus track fuel economy test.[11]

A second demonstration, sponsored by the Florida Department of Transportation and UMTA, is also ongoing. Its purpose is to determine the costs and benefits of retrofitting in-use General Motors 71-series diesel bus engines to methanol utilization. Three existing engines will be converted to methanol fueling in 1985 and put into service for approximately six months.[12]

Additional demonstrations, involving larger numbers of buses and the active interest and participation of individual transit authorities, are being planned. Seattle Metro Transit, one of the most stable and innovative transit authorities in the country, has already solicited bids for ten methanol buses and the Southern California Rapid Transit District (Los Angeles) is expected to seek bids for thirty methanol buses in the near future.[13] Other cities, such as New York and Denver, are also considering methanol demonstrations.[14] It is clear that both engine manufacturers and transit authorities are convinced that methanol buses provide a real alternative to diesel buses.

We recommend that Congress require, through UMTA, that all urban transit buses purchased in part with federal funds be methanol-fueled beginning in 1987. In addition, beginning in 1987 any engine retrofitting or any vehicle rehabilitation which involves significant engine overhauling and which involves federal subsidies should also utilize a methanol engine as the replacement engine.

In addition to the 80 percent share of capital expenses already available, the federal government would also provide 100 percent of the incremental cost of the methanol bus over and above the cost of a similar diesel bus. UMTA would also reimburse transit authorities for any incremental operating expenses due to the differential costs of methanol and diesel fuel.

### Bus Sales and Fuel Demand

A federal requirement that all new UMTA-subsidized urban buses be methanol-fueled would result in the purchase of three to five thousand new methanol buses each year. Since engine retrofits and major engine overhauls would also involve the installation of methanol engines, we would expect nearly the entire national urban bus fleet to be methanol-fueled by 1996. Including retrofits, there would be demand for approximately five thousand methanol bus engines per year. This demand should not be a problem for the bus engine manufacturers. The possibility of a market opening up for a new bus engine design should spur research and development efforts by the bus engine manufacturers, and results from this research could also be utilized in the design of methanol engines for interstate trucks, potentially a much bigger market than urban buses.

The initial impact of an urban methanol bus program on methanol consumption would be rather small, especially given the current surplus of natural gas–based methanol. If five thousand methanol engines were purchased for new vehicles and retrofits in 1987, for example, methanol consumption by those engines would only be about 7,200 BPD for 1987. Even if the entire urban bus fleet were methanol-fueled in 1996, total methanol consumed by buses would be just 72,000 BPD. The annual methanol bus fuel consumption is summarized in table 27. While this fuel demand is small, it is nevertheless

TABLE 27. Projected Methanol Bus and Fuel Demands from 1987 to 1996

| Year | Methanol Percentage of Urban Bus Fleet | Number of Methanol Buses | Methanol Consumption (BPD) |
|---|---|---|---|
| 1987 | 10 | 5,000 | 7,200 |
| 1988 | 20 | 10,000 | 14,000 |
| 1989 | 30 | 15,000 | 22,000 |
| 1990 | 40 | 20,000 | 29,000 |
| 1991 | 50 | 25,000 | 36,000 |
| 1992 | 60 | 30,000 | 43,000 |
| 1993 | 70 | 35,000 | 50,000 |
| 1994 | 80 | 40,000 | 58,000 |
| 1995 | 90 | 45,000 | 65,000 |
| 1996 | 100 | 50,000 | 72,000 |

important, especially in the 1987 to 1989 time period when only urban buses and federal fleet vehicles would be methanol-fueled. These fuel demands will be the first pure methanol demands for transportation and could act as catalysts to encourage investments by the energy and automotive industries that are necessary for major industry expansion.

One significant impact of an urban methanol bus program would be the introduction to the general public of methanol as a viable alternative transportation fuel. The use of methanol in transit buses would give methanol considerable public visibility and would enable citizens to directly compare its performance with that of diesel buses. The buses could be marked as methanol buses, furthering the public's identification of them. The fact that the federal government would be requiring methanol bus purchases would be a strong signal that it was serious about methanol as a future transportation fuel. This would likely spur both investor and citizen interest in the methanol implementation program.

### Improved Urban Air Quality

The use of methanol in urban transit buses nationwide should result in significant air quality improvements. Nearly all buses currently contain diesel engines, which emit high levels of several pollutants such as particulate matter and nitrogen oxides. Diesel buses also produce smoke and odor, which are irritants to most people, reactive hydrocarbons, and small amounts of unregulated pollutants such as sulfur oxides.

Because there is not as yet a large emission data base for methanol buses, it is instructive to consider first the emission products of methanol combustion from a theoretical perspective. Because of its physical and chemical composition, the combustion of methanol would not be expected to result in particulate matter, smoke, or sulfur oxides emissions. Methanol would also be expected to produce much lower emissions of nitrogen oxides and reactive hydrocarbons. Carbon monoxide levels from methanol engines would be expected to be similar to those from diesel engines. The two pollutants which would be expected to be emitted in greater amounts by methanol vehicles would be unburned methanol and formaldehyde. One very important difference between methanol and diesel combustion is that catalytic converters should be feasible on methanol vehicles, which will reduce emissions of reactive hydrocarbons, carbon monoxide, methanol, and formaldehyde. Catalysts are not used on diesel engines because of potential plugging by particulate emissions. Even with a catalyst, methanol and possibly formaldehyde emissions would be expected to be higher than from a diesel engine, but reactive hydrocarbon emissions would be expected to be much lower. Because methanol itself is considered to have a low photochemical reactivity, it would be expected that methanol buses would contribute less to photochemical oxidant (smog) levels in urban areas than do diesel buses.

Testing of heavy-duty bus or truck emissions is done by two different methods: chassis testing, where the bus or truck itself is tested on a large dynamometer, resulting in emissions measurements on a grams per mile basis; and engine testing, where the engine is pulled from the vehicle and the results are in grams per horsepower-hour.(A horsepower-hour is a unit of measure for work, in this case, a measure of the useful work of an engine.) Chassis testing is preferable in that it yields the actual grams per mile emission factors that are necessary for ambient air quality modeling. But because of the high cost of large chassis dynamometers, few laboratories are able to test large buses and trucks in this way. Therefore, most bus and truck emission testing is performed on engines, which require much smaller dynamometers. It is then necessary to apply a conversion factor to convert engine emissions in grams per horsepower-hour to vehicle emissions in grams per mile.

All methanol bus data to date were generated by engine testing. EPA has performed a comprehensive emission characterization of the M.A.N. methanol bus engine at Southwest Research Institute.[15] General Motors has reported the results of an internal test program with its methanol-fueled 6V-92TA bus engine.[16] These data are summarized in table 28, along with values typical for new diesel bus engines.[17] The first two columns in table 28 are directly comparable, since they both include data generated over the EPA transient engine test, which involves operating an engine over a simulated cycle that consists of constantly changing engine speed and load conditions. The General Motors data were collected from steady-state engine testing that operates an engine only at certain constant speed and load conditions. It can be seen that the preliminary data generally confirm the theoretical expecta-

TABLE 28. Diesel and Methanol Zero-Mile Bus Engine Emissions (Grams per Horsepower-Hour)

| Pollutant | Typical New Diesel Engines | New M.A.N. Methanol Engine with Catalyst | New General Motors Methanol Engine without Catalyst |
|---|---|---|---|
| Particulate | 0.57 | 0.04 | 0.17 |
| Oxides of nitrogen | 6.25 | 6.60 | 2.20 |
| Carbon monoxide | 3.22 | 0.31 | — |
| Total organics | 1.61 | 0.68 | 1.28 |
| Hydrocarbons | 1.51 | 0.001 | — |
| Methanol | 0 | 0.68 | 1.13 |
| Aldehydes | 0.10 | 0.001 | 0.15 |

*Note:* The first two columns of data were generated over the EPA heavy-duty engine transient test cycle while the results from the third column are from the thirteen-mode steady-state engine cycle.

tions. The M.A.N. engine, which included a catalytic converter, produced negligible particulate, hydrocarbon, and aldehyde emissions, and very low carbon monoxide emissions. Its nitrogen oxide emission level was comparable to typical diesel levels, and higher than expected. Of course, unburned methanol levels were higher. The General Motors engine, which did not involve catalytic aftertreatment, also produced promising results: very low particulate and nitrogen oxide emissions, and an aldehyde level only slightly higher than typical diesel levels. A catalyst would reduce aldehyde emissions considerably.

It is possible to estimate the value of the particulate and nitrogen oxides emission reductions that would result from the substitution of a fleet of methanol urban buses for diesel buses. This is only a preliminary analysis, as we do not have solid information on the future in-use emission levels of either methanol or diesel buses. As noted earlier, it is necessary first to derive gram per mile emission factors for diesel and methanol buses. EPA has recently tested seven in-use diesel buses, in their chassis configurations, in two different test programs. On average, these buses emitted 5.52 grams per mile particulate and 26.1 grams per mile nitrogen oxides.[18] Based on the emission data in table 28, as well as results from methanol test programs involving other types of diesel engines, it is fair to assume that methanol buses would reduce particulate levels by 90 percent and nitrogen oxide levels by 50 percent compared to diesel buses. This would result in projected methanol bus emission factors of 0.55 grams per mile particulate and 13.1 grams per mile nitrogen oxides, and emission reductions of 4.97 grams per mile particulate and 13.0 grams per mile nitrogen oxides. Since urban buses travel approximately 1.67 billion miles per year, the annual aggregate emission reductions from methanol buses would total 9,100 tons of particulate and 24,000 tons of nitrogen oxides.

All of the emissions from urban buses are emitted at ground level in urban areas near where people live, work, and travel, and thus have a high relative public exposure (especially compared to pollution from utility power plants, which are usually located in rural areas and which have very tall smokestacks). Based on a recent EPA comparison of various pollution control strategies, reasonable values for the cost-effectiveness of urban particulate and nitrogen oxides control are $10,000 per ton and $1,000 per ton, respectively.[19] Multiplying these emission reductions and cost-effectiveness values together results in savings of $91 million due to particulate control and $24 million due to nitrogen oxides control. Thus, the lower particulate and nitrogen oxides emissions alone from methanol urban buses would be worth approximately $115 million per year. This analysis must be viewed as preliminary, but clearly indicates that the urban emission reductions available from methanol buses are very valuable indeed.

There would be several other air quality benefits associated with the substitution of methanol for diesel fuel in urban buses which cannot be monetized. The organic (fuel-related) emissions in methanol vehicle exhaust are considered to be less photochemically reactive than organic emissions in diesel exhaust. Thus, methanol buses would reduce the photochemical reactivities of urban atmospheres and lower ozone levels. Methanol buses would also emit lower levels of sulfur oxides and other unregulated pollutants. Finally, diesel particulate samples have exhibited mutagenic and/or carcinogenic responses on various bacterial and animal tests. Although impossible to quantify at this time, the large reductions in diesel particulate emissions would likely lower the carcinogenic potencies of urban atmospheres.

In summary, the use of methanol in urban buses would result in significant reductions of particulate matter and oxides of nitrogen emissions in urban areas. Emissions of other pollutants, such as smoke and sulfur oxides, would also be reduced. While methanol buses will emit higher levels of methanol and formaldehyde, catalytic converters will ameliorate these problems and because of lower hydrocarbon levels the total photochemical reactivity impact of methanol buses will likely be less than that of diesel buses. The value of the emission reductions which can be monetized would likely exceed $100 million per year.

### Impacts on Transit Operators

We believe transit operators would welcome this program. There should be no major economic cost to the transit operators since the federal government would cover all incremental costs due to higher capital or operating costs. The local transit authorities would be responsible for providing methanol storage and handling capability at fueling sites. At first this might simply involve the transferring of one diesel tank to methanol, but ultimately it would require the installation of new, larger tanks. In chapter 7 we estimated the total cost of adding a new methanol fueling facility to be $28,000, assuming that existing storage tanks could not be used. Transit authorities in large cities would have to add several such facilities, but even then the cost to any individual authority would be relatively small. There would be a "learning curve" for transit personnel in adapting to a new fuel and different engine design, with implications for worker safety and maintenance, but both of these concerns are minimized by centralized fueling and maintenance systems.

The overriding impacts on transit operators would be favorable. One important benefit to transit operators would be the guarantee of a stable and secure fuel supply. As discussed above, diesel fuel supplies have become tighter in recent years, and the trend is expected to continue as diesel replaces gasoline in more automotive applications. In addition, with one-third of our crude oil imported, there continues to be the possibility of a crude oil supply disruption. Methanol would be produced from domestic feedstocks and could provide a secure supply for transit operators. Another advantage of methanol usage in urban buses would be the positive public relations associated with the program. Concerns over pollution, smoke, odor, and noise are barriers to better public support of urban buses. Local transit operators would be able to proclaim a "new" era in public transportation with "old, dirty" buses being replaced with new, clean buses. Furthermore, operators could emphasize that methanol is the "fuel of the future" and that their buses were on the leading edge of a new wave of methanol-fueled vehicles. Instead of always being seen as mundane and obsolete, buses might be seen as exciting and innovative.

### Impacts on the Federal Treasury

The federal government would reimburse transit authorities for any increases in cost of purchasing or operating a methanol bus as compared to purchasing or operating a diesel bus. It is not possible at this time to project the exact cost of a methanol bus, especially at first when manufacturers will want to recover development costs from initial sales. The only significant hardware-related incremental costs for methanol buses would be associated with larger on-board fuel tanks and catalytic converters. General Motors has stated in congressional testimony that at annual sales of two hundred fifty to three hundred methanol buses or more per year, methanol buses would likely cost between $6,000 and $7,000 more than comparable diesel buses.[20] Assuming annual sales of five thousand new buses or new bus engines per year, the increased federal outlays would be approximately $35 million per year for the early years of the program.

By 1992, however, we would expect that manufacturers would have recouped their original development costs and that the only incremental costs for methanol buses would be the hardware-related costs of larger fuel tanks and catalytic converters. At that time we would expect methanol buses to be only about $2,000 more expensive than their diesel counterparts. With sales of five thousand buses and bus engines per year, this would result in an added expense of $10 million per year. It is expected that there will be savings in maintenance costs for operators of methanol buses. Due to the lack of particulate emissions, there should be less engine wear, longer oil change intervals, etc. These maintenance savings should continue throughout the lifetime of the bus, and will help offset the higher capital costs. In fact, by 1992 we would expect that these maintenance savings for all of the methanol buses in operation by then would offset the slightly higher capital costs for the five thousand new buses purchased each year. Thus, we believe there would be no need for any incremental federal capital subsidies for methanol bus purchases after 1991.

Projecting the relative fuel costs of diesel and methanol urban buses is much more difficult, since the future costs of diesel and methanol fuels could vary considerably. The primary determinant of future diesel fuel prices will be future world oil prices. In view of the fact that world oil prices have risen from less than $3 per barrel in 1972 to nearly $40 per barrel in 1981, and have fallen to under $30 per barrel, it is clearly difficult to project future world oil prices with any confidence. Future methanol prices in the long run are expected to depend primarily on the production economics of large coal-to-methanol plants. Though there are no technical barriers to building such plants, no plants exist today and thus accurate costs are difficult to precisely estimate. Comparisons between diesel and methanol fuels are also complicated by the fact that both are in surplus as of 1984 and thus prices are relatively depressed. We will utilize both low and high price estimates for both diesel and methanol fuels, which will allow us to bracket a wide range of possible operating costs associated with both fuels.

Some bus transit authorities utilize No. 1 diesel fuel, others utilize No. 2 diesel, and still others use a combination of both during a calendar year. There is no good data on the actual percentages of No. 1 and No. 2 diesel fuels used by transit companies, so we will assume that the ratio is approximately fifty-fifty. DOE data indicate that the prices to end-users for No. 1 diesel and No. 2 diesel for the first ten months of 1984 were $0.93 and $0.83 per gallon, respectively (transit authorities do not pay fuel taxes).[21] Therefore, the average diesel fuel price to transit authorities is approximately $0.88 per gallon. Since it is unlikely that the world oil price in the late 1980s will be less than it is today, we will use this value as our low estimate for diesel fuel price. For our high diesel fuel price estimate, we will assume that our $0.40 per gallon highway fuel tax will cap U.S. oil import prices at $30 per barrel and retail diesel fuel prices, excluding taxes (urban buses would not be subject to the highway fuel tax either), at $1.00 per gallon. Thus, we will use $0.88 per gallon and $1.00 per gallon as our low and high costs of diesel fuel to transit authorities in the late 1980s and early 1990s.

Currently the methanol market is depressed, and methanol spot prices in the Gulf Coast are approximately $0.40 per gallon.[22] Adding $0.10 per gallon for distribution to northern cities would result in methanol delivered to transit authorities today for approximately $0.50 per gallon. Despite improving economic conditions, methanol prices may remain low because of large methanol plant capacity additions which have come on line recently or which will begin operation soon. For our low methanol price estimate, then, we will assume that the natural gas–based methanol surplus continues to hold delivered methanol prices down around $0.50 per gallon. As stated in chapter 5, the highway fuel tax would apply to all fossil fuel–derived highway fuels imported into the United States, including imported methanol. This was done to ensure that domestic coal-based methanol would be more competitive than imported methanol for most vehicle applications. Since urban buses are exempt from federal fuel taxes, however, transit operators would not be subject to the highway fuel tax on imported methanol. Industry sources have projected that imported methanol would cost approximately $0.53 per gallon through the early 1990s.[23] Adding $0.10 per gallon for distribution would yield a high methanol cost to transit authorities of $0.63 per gallon.

Methanol's energy content, 56,600 Btu/gallon, is only 44 percent of diesel's energy content of 128,000 Btu/gallon. Assuming similar energy efficiencies, an urban methanol bus would require 2.28 times as many gallons of fuel as a diesel bus. The entire fleet of urban buses in service in 1982 consumed 483 million gallons of fuel. We will assume that this overall fuel consumption will remain fairly constant into the 1990s if diesel fuel is used. For a baseline estimate of diesel fuel consumption for the first year of the methanol bus program we will assume that one-twelfth of the fleet would be replaced with new buses (or new bus engines) and that this fraction of the fleet would use approximately 10 percent of the overall diesel fuel consumption, or 48.3 million gallons. The corresponding methanol fuel consumption values would be 2.28 times greater, or 110 million gallons of methanol for the first year of the program and 1,100 million gallons of methanol for the entire fleet.

Using these fuel consumption figures and high and low estimates of fuel cost, table 29 gives the incremental costs associated with methanol fueling under the various pricing scenarios. It can be seen that methanol fuel could cost from $7 million to $27 million more than diesel fuel in the first year of a methanol urban bus program. Similarly, if the entire fleet was fueled by

TABLE 29. Projected First Year and Fleet Incremental Cost of Methanol Fuel to Transit Authorities

| Scenario | Diesel Price ($ per gallon) | Methanol Price ($ per gallon) | First Year Incremental Cost ($ million) | Entire Fleet Incremental Cost ($ million) |
|---|---|---|---|---|
| Low diesel cost/ low methanol cost | 0.88 | 0.50 | 12 | 120 |
| Low diesel cost/ high methanol cost | 0.88 | 0.63 | 27 | 270 |
| High diesel cose/ low methanol cost | 1.00 | 0.50 | 7 | 70 |
| High diesel cost/ high methanol cost | 1.00 | 0.63 | 21 | 210 |

methanol there would be an incremental cost of $70 million to $270 million per year. If one considers only the low diesel cost/low methanol cost and high diesel cost/high methanol cost scenarios, i.e., disregarding the two extreme scenarios, the incremental costs to the federal treasury would be $12 million to $21 million in the first year and $120 million to $210 million per year when the entire fleet was converted to methanol. This is our best estimate of the ranges for the incremental cost of methanol fuel.

Table 30 estimates the total incremental costs, both capital and operating, to the federal government for the methanol urban bus program for the years 1987 through 1996. As discussed earlier, we project annual capital expenditures to be $35 million greater for methanol buses in the early years of the program, with much of the increase due to research and development amortization. By 1992, we expect the capital costs to be only slightly higher and that these will be offset by maintenance savings for all methanol buses then in use. The incremental fuel costs in table 30 are the ranges of "best estimate" values in table 29, i.e., the values determined from our analyses of the low diesel cost/low methanol cost and high diesel cost/high methanol cost fuel price scenarios. It can be seen from table 30 that the incremental capital and fuel costs are comparable in the early years of the program but that by 1990 incremental fuel costs begin to dominate. The total annual federal subsidy would be between $47 million and $56 million in 1987 and would grow to between $120 million and $210 million by 1996 when all urban buses would be methanol-fueled.

In summary, under our proposal we would expect that the federal government would have to provide between $50 million and $200 million annually to local transit authorities for the methanol urban bus program through the mid-1990s. In the late 1980s these funds would be used to subsidize bus purchases and higher fuel costs but by the early 1990s they would be used

TABLE 30. Projected Total Annual Incremental Cost to the Federal Government of the Methanol Urban Bus Program ($ Million)

| Year | Incremental Capital Cost | Incremental Fuel Cost | Total Incremental Cost |
|---|---|---|---|
| 1987 | 35 | 12–21 | 47–56 |
| 1988 | 35 | 24–42 | 59–77 |
| 1989 | 35 | 36–63 | 71–98 |
| 1990 | 35 | 48–84 | 83–119 |
| 1991 | 35 | 60–105 | 95–140 |
| 1992 | 0 | 72–126 | 72–126 |
| 1993 | 0 | 84–147 | 84–147 |
| 1994 | 0 | 96–168 | 96–168 |
| 1995 | 0 | 108–89 | 108–89 |
| 1996 | 0 | 120–210 | 120–210 |

exclusively to cover incremental fuel costs. Compensating for these federal expenditures are the value of the reduced urban air pollution (previously estimated at over $100 million per year), a secure fuel supply for transit buses, an improved public image for mass transit, and the part this program could play in launching methanol as an alternative transportation fuel to imported petroleum. We believe the overall impact of a methanol urban bus program would be a significant net benefit for the American public.

## CHAPTER 10 Methanol Requirement for Federal Vehicles

The federal government vehicle fleet comprises a very small percentage of the national vehicle fleet. The federal government owns 430,600 vehicles, including 110,000 passenger cars, 264,400 light trucks, and 56,200 heavy trucks and miscellaneous vehicles.[1] There are approximately 127 million automobiles and light trucks now in use in the United States.[2] Including postal service and military light trucks, the federal government owns just 0.3 percent of all highway vehicles in use. Excluding postal service and military light trucks, which are rarely used for passenger transport, the federal government operates just 0.15 percent of highway passenger vehicles.

The General Services Administration calculated federal fleet fuel consumption to be 275 million gallons in fiscal year 1982.[3] This figure excludes agencies with fleets less than 2,000 vehicles, so it is a slight underestimate. This fuel consumption is equivalent to 18,000 barrels per day (BPD). Total U.S. passenger car and light truck fuel consumption totalled 6,100,000 BPD in 1982.[4] Thus, the federal fleet accounts for just 0.3 percent of all passenger car and light truck fuel consumption.

It is apparent that federal fleet utilization of methanol cannot be the cornerstone of a methanol implementation program. Even if the entire federal fleet operated on methanol the increase in demand for methanol would be only a small fraction of that necessary to spur significant methanol production capacity. Nevertheless, the use of methanol by the federal fleet can be an important *symbolic* step in an overall methanol implementation program. There is precedent for applying petroleum conservation measures to the federal fleet as an example for other fleet operators and individual citizens to follow; for example, Executive Order 12003, issued on July 20, 1977, mandated average fuel economy levels for federal fleet vehicle purchases.[5]

We recommend that, beginning in 1987, all federal fleet (civilian agencies, postal service, and military agencies) purchases of passenger cars and light trucks be exclusively methanol-fueled vehicles and that federal fueling facilities be modified to include methanol storage and dispensing capabilities. The only exceptions would be for those federal fleets that are very small and do not have their own fueling facilities, *and* are located in areas which do not have any public methanol service stations. Since most government vehicles are sold after five or six years of operation, this proposal would result in the federal fleet being composed almost exclusively of methanol vehicles by 1993. At that time, government fueling facilities would not need to maintain fuels other than methanol on site. Neither vehicle nor fuel supplies should constrain this proposal, as at least two vehicle manufacturers (Ford and Volkswagen) have "assembly-line produced" methanol vehicles, and it is expected that there will be an excess of methanol supply in the near future, even before start-up of any large coal-to-methanol plants.

### Increased Methanol Vehicle Sales and Fuel Demand

Enactment of this proposal would have several direct impacts on the methanol industry. Table 31 gives projected methanol fuel demand by the federal fleet based on the following assumptions: (1) 15 percent of the existing federal fleet of passenger cars and light trucks will be replaced by methanol vehicles during each year from 1987 to 1992 with nearly the entire fleet being methanol-fueled by 1993; (2) there is no growth in the federal fleet; and (3) methanol vehicles will have approximately 20 percent better energy efficiencies than the vehicles they will be replacing, so that they will require about 1.7 times as much methanol as petroleum on a volumetric basis. It can be seen that methanol demand would be 4,600 BPD in 1987, rising to 30,600 BPD in 1993. The maximum demand of 30,600 BPD is one-third of the output of a large coal-to-methanol plant. Implementation of this proposal would require the purchase of approximately 56,000 methanol-fueled passenger cars and light trucks per year beginning in 1987. The proposal would require federal fueling facilities to add methanol storage and dispensing capabilities, involving installation of storage tanks and pumps. This would encourage research and marketing of hardware. It would encourage the development of expertise

TABLE 31. Projected Methanol Fuel and Vehicle Demands by the Federal Fleet from 1987 to 1993

| Year | Methanol Percentage of Federal Fleet | Number of Methanol Vehicles | Methanol Consumption (BPD) |
|---|---|---|---|
| 1987 | 15 | 56,000 | 4,600 |
| 1988 | 30 | 112,000 | 9,200 |
| 1989 | 45 | 168,000 | 13,800 |
| 1990 | 60 | 225,000 | 18,400 |
| 1991 | 75 | 281,000 | 23,000 |
| 1992 | 90 | 337,000 | 27,500 |
| 1993 and later | 100 | 374,000 | 30,600 |

on methanol engine diagnosis and repair by federal fleet operators and mechanics responsible for engine and vehicle maintenance. Finally, the federal fleet would provide a very large data base to assess the overall performance of methanol vehicles with respect to durability, emissions, energy efficiency, and driveability.

In terms of direct fuel and vehicle demands, the federal fleet can play only a small role in the development of an overall methanol implementation plan. But the *indirect* benefits of methanol utilization in the federal fleet may well outweigh its direct impacts. With respect to the corporate investment process, the methanol fuel and vehicle demands, though small relative to ultimate production goals, are important because they will be among the initial demands and will encourage investments by the energy and automotive industries that will be necessary for industry expansion. Conversion of the federal fleet to methanol will also emphasize to investors that the federal government is serious about a methanol implementation program, which should alleviate the confusion which many planners have about government intentions with respect to alternative fuels.

Indirect benefits are also likely with respect to acceptance of methanol as a transportation fuel by the general public. Use of methanol by the government should cause many citizens to take note and to consider purchasing methanol vehicles for their own uses. Further, federal fleet vehicles are often marked as such and are thus very visible to the general public. This raises the possibility of some creative educational methods. Specific agencies which have their own vehicle pools could advertise particular advantages of methanol which are pertinent to their governmental responsibilities. For example, Department of Energy vehicles could have stickers pointing out that the fuel they are using came from domestic feedstocks, Environmental Protection Agency vehicles could proclaim the cleaner exhaust emissions, and the Department of Labor could point out that its fuel was produced by American labor. The postal service could even modify its slogan "neither rain, nor sleet, nor hail, nor snow" to include the phrase "nor imported oil cutoff"! We are not suggesting a large federal advertising campaign, only some ways in which the government could cheaply yet effectively educate the American public about the benefits of methanol fuel.

### Impacts on the Federal Treasury

A requirement that all 1987 and later model year federal fleet vehicles be methanol-fueled could affect both capital and operating costs for federal fleet operators. In the near term, before manufacturers can take full advantage of economies of scale, methanol vehicles will likely be slightly more expensive than similar gasoline-fueled vehicles. In recent congressional testimony, representatives of two domestic manufacturers stated that for production runs of "tens of thousands" of vehicles the methanol premium would likely be approximately $200 per car, while the premium should be zero for production volumes greater than 100,000 vehicles.[6] It is expected that large-scale methanol vehicle production would coincide with the implementation of the methanol vehicle tax credit program in 1990. After 1990, then, methanol vehicles should cost no more than similar gasoline vehicles. For the years 1987 through 1989, however, we will assume that methanol vehicles would cost $200 more than comparable gasoline vehicles. With projected federal fleet purchases of 56,000 vehicles per year, this would result in an incremental capital cost of $11 million per year from 1987 through 1989.

The major component of any change in operating costs will be fuel cost. Projecting future prices of gasoline and methanol fuels with any confidence is very difficult. The prices of both petroleum and methanol have been very volatile during the last decade. Both fuels are in surplus today and are available at depressed prices.

The average price of unleaded regular gasoline sold to bulk end-users in the United States during the first six months of 1984 was $0.92 per gallon.[7] This price excludes state and federal fuel excise taxes but includes the cost of delivering the fuel to the end-user.[8] We will assume that baseline U.S. oil prices will remain stable so that gasoline prices excluding the highway fuel tax will remain at $0.92 per gallon through 1989. By 1990, however, we would expect oil prices to rise to the $30 per barrel cap imposed by the highway fuel tax. This would result in baseline gasoline prices of $1.00 per gallon beginning in 1990. The new highway fuel tax would apply to gasoline purchased by federal fleet operators, just as the current federal fuel excise tax does.[9] The total gasoline cost for federal vehicles would be the sum of the baseline gasoline cost and the highway fuel tax. This cost would be $0.92 per gallon in 1987 (no tax) and would rise progressively to $1.40 per gallon in 1992 and later years (maximum $0.40 per gallon tax).

Current methanol prices in the Gulf Coast are approximately $0.40 per gallon.[10] Transporting and distributing the methanol to federal fueling facilities would likely add another $0.10 per gallon to the methanol price, resulting in a delivered price of $0.50 per gallon today. Due to the present methanol surplus, we will assume that $0.50 per gallon would be a reasonable methanol price until 1990 when large-scale demand for methanol would begin. We projected in chapter 5 that methanol could be produced from coal and distributed to localities for $0.71 per gallon. Thus, we will assume that methanol will cost $0.50 per gallon through 1989 and $0.71 per gallon thereafter.

Gasoline has an energy content of approximately 116,000 Btu/gallon. Methanol has an energy content of 56,600 Btu/gallon, just 49 percent of the value for gasoline. As noted earlier, however, it is expected that methanol vehicles will be approximately 20 percent more efficient, on a Btu basis, than the gasoline vehicles they will be replacing. Thus, a methanol vehicle would consume about 1.7 times as many gallons of fuel as a similar gasoline vehicle. Federal fleet fuel consumption totalled 275 million gallons in 1982. We will assume that overall federal fleet fuel consumption would remain constant into the 1990s if gasoline were to continue to be the sole fuel. We will assume that new vehicles will comprise 15 percent of the federal fleet in 1987 and will consume 15 percent of the total fuel. If these were gasoline vehicles, this would amount to 41 million gallons. The corresponding methanol fuel consumption values would be 1.7 times greater, or 70 million gallons for the first year of the program and 468 million gallons if the entire fleet were methanol-fueled.

Table 32 estimates the total incremental costs/savings, both capital and operating, to the federal government for the methanol federal fleet requirement. As discussed earlier, we project annual capital expenditures to be $11 million per year greater for methanol cars and light trucks in the first three years of the program, but by 1990 it is expected that methanol vehicles will be no more expensive than comparable gasoline vehicles. The incremental fuel savings are based on the fuel cost and consumption projections developed in this chapter and assume that 15 percent of the fleet would be converted to methanol each year beginning in 1987. It can be seen that methanol fueling is projected to result in fuel savings each year, with the magnitude of the savings growing in the early 1990s due to the increasing highway fuel tax. In 1987, the first year of the program, methanol would increase total federal fleet costs by $8 million. In 1988 the methanol fuel savings would equal the increased capital cost. By 1989 the fuel savings would exceed the capital costs by $17 million, and the overall savings would increase to $53 million by 1993 when the entire fleet would be methanol-fueled.

TABLE 32. Projected Total Annual Incremental Cost to the Federal Government of the Methanol Federal Fleet Requirement ($ Million)

| Year | Incremental Capital Cost | Incremental Fuel Cost | Total Incremental Cost |
|---|---|---|---|
| 1987 | 11 | −3 | 8 |
| 1988 | 11 | −12 | −1 |
| 1989 | 11 | −28 | −17 |
| 1990 | 0 | −5 | −5 |
| 1991 | 0 | −23 | −23 |
| 1992 | 0 | −48 | −48 |
| 1993 and later | 0 | −53 | −53 |

*Note:* Negative values represent incremental savings for the methanol scenario.

In summary, a requirement that new federal fleet vehicles be methanol-fueled appears to be an excellent investment for the federal government. By 1993, when the entire federal fleet would be methanol-fueled, the savings are projected to be over $50 million per year. Even more important than the monetary savings, however, is that a methanol federal fleet requirement would be an important symbolic gesture that would emphasize to citizens and the private sector that the federal government is serious about launching the methanol transition.

# Notes

## Chapter 1

1. National Academy of Sciences, National Research Council, *Acid Deposition: Atmospheric Processes in Eastern North America* (Washington, D.C.: National Academy Press, 1983).

2. Ibid.

3. Environmental Protection Agency, Office of Research and Development, *The Acidic Deposition Phenomenon and Its Effects: Critical Assessment Review Papers,* vol. 1, EPA-600/8-83-016aF (September 1984).

4. Environmental Protection Agency, Office of Research and Development, *The Acidic Deposition Phenomenon and Its Effects: Critical Assessment Review Papers,* vol. 2, EPA-600/8-83-016bF (September 1984).

5. Ibid.

6. Ibid.

7. Ibid.

8. Ibid.

9. Ibid.

10. Ibid.

11. Ibid.

12. Ibid.

13. *The Acidic Deposition Phenomenon,* vol. 1.

14. Ibid.

15. Ibid.

16. Ibid.

17. *Acid Deposition,* p. 7.

18. Ibid.

19. "Final Study by White House Panel Urges 'Meaningful' Sulfur, Nitrogen Emission Cuts," *Environment Reporter,* September 14, 1984, p. 758.

20. Office of Technology Assessment, *Acid Rain and Transported Air Pollutants—Implications for Public Policy,* OTA-0-204 (Washington, D.C.: Office of Technology Assessment, June 1984); "Pourtney: Flexible Acid Rain Plan Would Save over $1-Billion, Create Jobs," *Inside E.P.A.,* December 2, 1983.

21. DOE Energy Information Administration, "1982 Annual Energy Review," April 1983.

22. Ibid.; DOE Energy Information Administration, ''Monthly Energy Review,'' DOE/EIA-0035 (85/01), January 1985.

23. ''1982 Annual Energy Review.''

24. ''Geological Estimates of Undiscovered Recoverable Conventional Resources of Oil and Gas in the United States, A Summary,'' U.S. Geological Survey Circular 860, 1981; U.S. Geological Survey Open-File Report 83-728, 1983.

25. ''Monthly Energy Review,'' January 1985.

26. DOE Office of Policy, Planning, and Analysis, ''Energy Projections to the Year 2010,'' DOE/PE-0029/2, October 1983; Office of Technology Assessment, ''Increased Automobile Fuel Efficiency and Synthetic Fuels—Alternatives for Reducing Oil Imports,'' OTA-E-185, September 1982; Exxon Company, USA, ''Energy Outlook 1980–2000,'' December 1980; Conoco, Inc., ''World Energy Outlook through 2000,'' April 1984.

27. ''Energy Projections to the Year 2010.''

28. Council of Economic Advisers, ''Economic Indicators,'' March 1985.

29. Robert H. Williams, ''A $2 a Gallon Political Opportunity,'' in *The Dependence Dilemma: Gasoline Consumption and America's Security,* edited by Daniel Yergin (Cambridge, Mass.: Harvard University Center for International Affairs, 1980); American Petroleum Institute, ''The Social Costs of Incremental Oil Imports: A Survey and Critique of Present Estimates,'' Discussion Paper No. 025, February 1982.

30. ''Increased Automobile Fuel Efficiency and Synthetic Fuels,'' p. 69.

31. ''1982 Annual Energy Review.''

## Chapter 2

1. DOE Energy Information Administration, ''1982 Annual Energy Review,'' April 1983.

2. Ibid.

3. Sam H. Schurr et al., *Energy in America's Future: The Choices before Us* (Baltimore: Johns Hopkins University Press for Resources for the Future, 1979).

4. ICF, Inc., ''Summary of Acid Rain Analyses Undertaken by ICF for the Edison Electric Institute, National Wildlife Federation, and Environmental Protection Agency,'' May 1982; United Mine Workers of America, ''Employment Impacts of Acid Rain,'' June 1983.

5. DOE Office of Policy, Planning, and Analysis, ''Energy Projections to the Year 2010,'' DOE/PE-0029/2, October 1983.

6. EPA Office of Mobile Sources, ''Preliminary Perspective on Pure Methanol Fuel for Transportation,'' EPA 460/3-83-003, September 1982.

7. Office of Technology Assessment, ''Increased Automobile Fuel Efficiency and Synthetic Fuels—Alternatives for Reducing Oil Imports,'' OTA-E-185, September 1982.

8. Southwest Research Institute, ''Characterization of Exhaust Emissions from Methanol- and Gasoline-Fueled Automobiles,'' EPA 460/3-82-004, August 1982.

9. Systems Applications, Inc., ''Impact of Methanol on Smog: A Preliminary Estimate,'' prepared for ARCO Petroleum Products Co., February 1983; Jet Propul-

sion Laboratory and California Institute of Technology, "California Methanol Assessment," vol. 2, prepared for the Electric Power Research Institute and the California Energy Commission, March 1983, chap. 6.

10. EPA Office of Air Quality Planning and Standards, "National Air Quality and Emission Trends Report, 1983," EPA 450/4-84-029, April 1984.

11. Southwest Research Institute, "Emission Characterization of a Spark-Ignited, Heavy-Duty, Direct-Injected Methanol Engine," EPA 460/3-82-003, November 1982; R. R. Toepel, J. E. Bennethum, and R. E. Heruth, "Development of Detroit Diesel Allison 6V-92TA Methanol-Fueled Coach Engine," Society of Automotive Engineers Paper Number 831744.

12. Craig Harvey, "Determination of a Range of Concern for Mobile Source Emissions of Methanol," Internal Staff Report, EPA-AA-TSS-83-6, July 1983; Penny Carey, "Determination of a Range of Concern for Mobile Source Emissions of Formaldehyde Based Only on Its Toxicological Properties," Internal Staff Report, EPA-AA-TSS-83-5, July 1983.

## Chapter 3

1. Statement of Ford Motor Company before the House Subcommittee on Fossil and Synthetic Fuels, September 24, 1982.

2. A. J. Sobey, "Energy Situation and the Automobile Industry," presentation to the Commonwealth Club of California, March 14, 1983.

3. General Accounting Office, "Removing Barriers to the Market Penetration of Methanol Fuels," GAO/RCED-84-36, October 27, 1983.

4. *Federal Register* 49 (April 10, 1984): 14244.

5. "Ford: Methanol Test Fleets Won't Launch the Fuel in America," *Alcohol Week,* May 28, 1984.

## Chapter 4

1. Jack Faucett Associates, Inc., "Employment Associated with a Domestic Methanol Fuel Production Industry," prepared for EPA Office of Mobile Sources, April 1985.

2. Jack Faucett Associates, Inc., "Comparison of the Macroeconomic Impacts of Highway Fuel Taxes," prepared for EPA Office of Mobile Sources, May 1985.

3. Ibid.

4. Ibid.

5. "Employment Associated with a Domestic Methanol Fuel Production Industry."

6. "Conservation and Alternative Fuels in the Transportation Sector," Report of the Transportation Task Force of the Solar Energy Research Institute Solar/Conservation Study, June 25, 1980; Charles L. Gray, Jr., and Frank von Hippel, "The Fuel Economy of Light Vehicles," *Scientific American* 244 (May 1981): 48; Richard Shackson and H. James Leach, "Maintaining Automotive Mobility: Using Fuel Economy and Synthetic Fuels to Compete with OPEC Oil," Interim Report, Mellon Institute, August 18, 1980.

7. "Macroeconomic Impacts."

8. ICF, Inc., "Summary of Acid Rain Analyses Undertaken by ICF for the Edison Electric Institute, National Wildlife Federation, and Environmental Protection Agency," May 1982; United Mine Workers of America, "Employment Impacts of Acid Rain," June 1983.

9. Gregg Marland, "Carbon Dioxide Emission Rates for Conventional and Synthetic Fuels," *Energy, The International Journal* 8 (1983): 981–92.

## Chapter 5

1. Office of Technology Assessment, "Increased Automobile Fuel Efficiency and Synthetic Fuels—Alternatives for Reducing Oil Imports," OTA-E-185, September 1982.

2. Robert H. Williams, "A $2 a Gallon Political Opportunity," in *The Dependence Dilemma: Gasoline Consumption and America's Security,* edited by Daniel Yergin (Cambridge, Mass.: Harvard University Center for International Affairs, 1980).

3. Frank von Hippel and Robert H. Williams, "Toward an Automotive Energy Policy," Draft report, March 1982.

4. American Petroleum Institute, "The Social Costs of Incremental Oil Imports: A Survey and Critique of Present Estimates," Discussion Paper No. 025, February 1982.

5. "Increased Automobile Fuel Efficiency and Synthetic Fuels," p. 69.

6. EPA Office of Mobile Sources, "Preliminary Perspective on Pure Methanol Fuel for Transportation," EPA 460/3-83-003, September 1982.

7. "Increased Automotive Fuel Efficiency and Synthetic Fuels."

8. DOE Energy Information Administration, "Price Elasticities of Demand for Motor Gasoline and Other Petroleum Products," May 1981.

9. Dave Bodde and Philip Webre, "Oil Import Fees," Congressional Budget Office internal memorandum to Ray Scheppach, March 22, 1982.

10. DOE Energy Information Administration, "Monthly Energy Review," DOE/EIA-0035 (84/01), January 1984.

11. Ibid.

12. DOE Energy Information Administration, "1982 Annual Energy Review," April 1983.

13. Joel B. Smith, "Trends in Energy Use and Fuel Efficiency in the U.S. Commercial Airline Industry," *Transportation Research Record* 870 (1982): 91.

14. Ibid.

15. "Conservation and Alternative Fuels in the Transportation Sector," Report of the Transportation Task Force of the Solar Energy Research Institute Solar/Conservation Study, June 25, 1980.

16. G. Kulp and M. C. Holcomb, *Transportation Energy Data Book,* 6th ed., DOE/ORNL-5883, 1982.

17. *The Dependence Dilemma: Gasoline Consumption and America's Security,* edited by Daniel Yergin (Cambridge, Mass.: Harvard University Center for International Affairs, 1980).

18. Jack Faucett Associates, Inc., "Comparison of the Macroeconomic Impacts of Highway Fuel Taxes," prepared for EPA Office of Mobile Sources, May 1985.
19. Ibid.
20. "Price Elasticities."
21. "Macroeconomic Impacts."
22. Ibid.
23. Ibid.
24. Ibid.
25. *Transportation Energy Data Book.*
26. Ibid.
27. "Conservation and Alternative Fuels."
28. "Energy Tax Options," Hearing before the Senate Subcommittee on Energy and Agricultural Taxation of the Committee on Finance, June 9, 1982.

## Chapter 6

1. U.S. Senate, Committee on the Budget, Subcommittee on Synthetic Fuels, "Synthetic Fuels," September 27, 1979.
2. "Energy Security Act," Public Law 96-294, June 30, 1980.
3. Booz-Allen and Hamilton, Inc., "Analysis of Economic Incentives to Stimulate a Synthetic Fuels Industry," August 27, 1979.
4. EPA Office of Mobile Sources, "Preliminary Perspective on Pure Methanol Fuel for Transportation," EPA 460/3-83-003, September 1982.
5. DOE Office of Policy, Planning, and Analysis, "Energy Projections to the Year 2010," DOE/PE-0029/2, October 1983.
6. "Preliminary Perspective"; Office of Technology Assessment, "Increased Automobile Fuel Efficiency and Synthetic Fuels—Alternatives for Reducing Oil Imports," OTA-E-185, September 1982.
7. "Preliminary Perspective."
8. "Surface Transportation Assistance Act of 1982," Public Law 97-424, January 6, 1983.
9. Jack Faucett Associates and Battelle Columbus Laboratories, *Energy and Precious Fuels Requirements of Fuel Alcohol Production,* vol. 4, *Appendices G and H: Methanol from Coal,* prepared for DOE and NASA, December 1982.
10. "Energy Projections to the Year 2010."
11. Ibid.

## Chapter 7

1. Department of Commerce, Bureau of the Census, "1977 Census of Retail Trade, Establishment and Firm Size," March 1980.
2. Department of Commerce, "Franchising in the Economy, 1981–1983," January 1983.
3. "At Last, a Comprehensive Count," *Lundberg Letter,* August 5, 1983.
4. Personal communication with a representative of the California Energy Commission, June 1984.

5. *Federal Register* 38 (January 10, 1973): 1254.

6. *Federal Register* 39 (May 7, 1974): 16123.

7. *Federal Reporter,* 2d ser., 501 (1974): 722.

8. 1982 *National Petroleum News* Factbook Issue.

9. Pacific Environmental Services, Inc., "Evaluation of Air Pollution Control Alternatives for Gasoline Marketing Industry," draft for EPA Office of Air Quality Planning and Standards, March 1984.

10. Personal communication with a representative of the California Energy Commission, June 1984.

11. R. Dwight Atkinson, "Distribution of Methanol as a Transportation Fuel," EPA Internal Staff Report, EPA-AA-SDSB-82-13, June 1982.

## Chapter 8

1. Jet Propulsion Laboratory and California Institute of Technology, "California Methanol Assessment," vol. 2, prepared for the Electric Power Research Institute and the California Energy Commission, March 1983.

2. Responses by Helen Petrauskas, Ford Motor Company, and Robert Frotsch, General Motors, to questions at the Joint Hearing by the Subcommittees on Fossil and Synthetic Fuels and Energy Conservation and Power on April 25, 1984.

## Chapter 9

1. Personal communication with a representative of the American Public Transit Association, February 1985.

2. American Public Transit Association, "Transit Fact Book, 1981," October 1981.

3. Personal communication with a representative of the Urban Mass Transportation Administration, January 1984.

4. "Transit Fact Book, 1981."

5. Booz-Allen and Hamilton, Inc., "Methanol Engine Conversion, Phase 1: Feasibility Study," prepared for Florida Department of Transportation, March 1983.

6. Southwest Research Institute, *Emissions from Heavy-Duty Engines Using the 1984 Transient Test Procedure,* vol. 2, *Diesel,* EPA-460/3-81-031, 1981.

7. *Federal Register* 50 (March 15, 1985): 10606.

8. Jeff Alson, "A Brief Summary of the Technical Feasibility, Emissions, and Fuel Economy of Pure Methanol Engines," EPA-AA-SDSB-82-1, December 1981.

9. A. Neitz and F. Chmela, "Results of Further Development in the M.A.N. Methanol Engine," paper presented at the Sixth International Symposium on Alcohol Fuels Technology, May 1984; H. K. Bergmann and K. D. Holloh, "Field Experience with Mercedes-Benz Methanol City Buses," paper presented at the Sixth International Symposium on Alcohol Fuels Technology, May 1984; R. R. Toepel, J. E. Bennethum, and R. E. Heruth, "Development of Detroit Diesel Allison 6V-92TA Methanol-Fueled Coach Engine," Society of Automotive Engineers Paper Number 831744.

10. M. D. Jackson, Roy A. Renner, and K. D. Smith, "California's Experience

with Methanol-Fueled Transit Buses,'' paper presented at the Sixth International Symposium on Alcohol Fuels Technology, May 1984.

11. Michael D. Jackson, Stefan Unnasch, Cindy Sullivan, and Roy A. Renner, ''Transit Bus Operation with Methanol Fuel,'' Society of Automotive Engineers Paper Number 850216.

12. ''Methanol Engine Conversion.''

13. *Alcohol Week,* November 12, 1984.

14. *Alcohol Week,* January 7, 1985.

15. Southwest Research Institute, ''Emission Characterization of a Spark-Ignited, Heavy-Duty, Direct-Injected Methanol Engine,'' EPA 460/3-82-003, November 1982.

16. ''Development of Detroit Diesel Allison 6V-92TA Methanol-Fueled Coach Engine.''

17. Richard D. Wilson and Charles Gray, Jr., ''The Environmental Impacts of Methanol as a Transportation Fuel,'' paper presented to the National Science Foundation Workshop on Automotive Use of Methanol-Based Fuels, January 10, 1985.

18. Southwest Research Institute, ''Characterization of Heavy-Duty Motor Vehicle Emissions under Transient Driving Conditions,'' NTIS Order Number PB 85-124 154, October 1984; Southwest Research Institute, ''Emissions Characterization of Heavy-Duty Diesel and Gasoline Engines and Vehicles,'' EPA 460/3-85-001, May 1985.

19. ''Regulatory Impact Analysis, Oxides of Nitrogen Pollutant Specific Study, and Summary and Analysis of Comments,'' Regulatory Support Document, EPA Office of Mobile Sources, March 1985.

20. Written response by General Motors to questions from the Joint Hearing before the Subcommittees on Fossil and Synthetic Fuels and Energy Conservation and Power on April 25, 1984.

21. DOE Energy Information Administration, ''Petroleum Marketing Monthly,'' DOE/EIA-0380 (84/10), October 1984.

22. *Alcohol Week,* December 24, 1984.

23. Letter from Harry W. Buchanan, Celanese Corporation, to Charles Gray, Environmental Protection Agency, February 7, 1984.

## Chapter 10

1. General Services Administration, ''Federal Motor Vehicle Fleet Report for Fiscal Year 1982,'' June 1983.

2. M. C. Holcomb and S. Koshy, *Transportation Energy Data Book,* 7th ed., DOE/ORNL-6050, June 1984.

3. ''Federal Motor Vehicle Fleet Report for Fiscal Year 1982.''

4. *Transportation Energy Data Book.*

5. ''Conservation and Alternative Fuels in the Transportation Sector,'' Report of the Transportation Task Force of the Solar Energy Research Institute Solar/Conservation Study, June 25, 1980, chap. 2.

6. Responses by Helen Petrauskas, Ford Motor Company, and Robert Frotsch,

General Motors, to questions at the Joint Hearing by the Subcommittees on Fossil and Synthetic Fuels and Energy Conservation and Power on April 25, 1984.

7. DOE Energy Information Administration, "Petroleum Marketing Monthly," DOE/EIA-0380 (84/10), October 1984.

8. Personal communication with a representative of the DOE National Energy Information Center, May 1984.

9. Personal communication with a representative of the Defense Logistics Agency, Department of Defense, February 1985.

10. *Alcohol Week,* December 24, 1984.